EVOLUTIONARY MEDICINE

EVOLUTIONARY MEDICINE

Stephen C. Stearns
YALE UNIVERSITY

Ruslan Medzhitov
YALE UNIVERSITY

Sinauer Associates, Inc. Publishers
Sunderland, Massachusetts U.S.A.

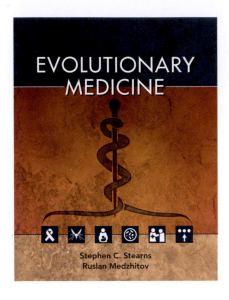

CANCER

REPRODUCTION

PATIENT

PATHOGEN

DISEASE

SELECTION

About the cover

The cover of *Evolutionary Medicine* was created by Joanne Delphia, who also designed the interior of the book.

Evolutionary Medicine

For information, address:
Sinauer Associates, Inc.
P. O. Box 407
Sunderland, MA 01375 U.S.A.
Fax: 413-549-1118
E-mail: publish@sinauer.com
Internet: www.sinauer.com

Library of Congress Cataloging-in-Publication Data
Stearns, S. C. (Stephen C.), 1946-
   Evolutionary medicine / Stephen C. Stearns, Yale University, Ruslan Medzhitov, Yale University.
      pages cm
   ISBN 978-1-60535-260-2
   1. Diseases--Causes and theories of causation. 2. Human evolution. 3. Environmentally induced diseases. I. Medzhitov, Ruslan. II. Title.
   RB152.S75 2016
   616.07'1--dc23

                          2015015915

Printed in China
5 4 3 2 1

To my grandchildren
S. C. S.

To my parents
R. M.

Brief Contents

Contents

3   What is a Disease?   87

4   Defenses   101

7   Reproductive Medicine   191

8 Mismatch 219

11 Open Questions and Other Issues   267

Preface

Interest in evolutionary insights into medical and public health issues has been growing rapidly in colleges, universities, and schools of medicine and public health. Some of these insights have immediate practical relevance: they can reduce suffering and save lives. Others are foundational: they help to explain the basic natures of patients and diseases. Both types of insights are important; both are introduced here. We wrote this book because we have both enjoyed teaching students about these insights and wanted to share them with a broader audience.

Evolutionary biology, like physics and chemistry, is a basic science that underpins all of biology. It is not a specialty like genetics, biochemistry, development, or physiology. Instead, it helps to explain both the origins and the current state of everything we find in organisms, including the subjects of such special fields. Evolution thus provides a framework that integrates medical knowledge. Its insights are distributed across many of the traditional medical and public health specialties, with particular emphasis on the practical consequences of genetic variation in disease resistance and drug metabolism, the evolution of resistance to antibiotics and chemotherapy, cancer as an evolutionary process, and all of the consequences of mismatches to modernity, particularly as they explain autoimmune diseases, allergies, obesity, and cardiovascular disease.

It also contains many examples of biology that are fascinating for their own sake, amazing and beautiful, and simply fun to think about.

Road map

The book begins in Chapter 1 with an introduction to evolutionary thinking that covers both current dynamic processes and the deep patterns of history and relationship. Chapter 2 then asks, What is a patient? It answers that question from a series of diverse perspectives each of which adds a layer of understanding. Chapter 3 asks the complementary question, What is a disease? How should we think about disease categories from an evolutionary and functional perspective? Where are the causes of diseases located? Why there and not elsewhere? Some of those causes are found in patients; others in pathogens; many in the interactions between them.

Causes of disease located in patients are further explored in Chapter 4, which discusses the nature of defenses, the strategies that determine how they are deployed, and the costs that they impose as well as the benefits that they yield.

Causes of disease located in pathogens are described in Chapter 5, which discusses the evolution of intrinsic virulence, how pathogens evade and manipulate host defenses, why and how they evolve resistance to treatment, and how therapy might be made evolution-proof.

Cancer is an evolutionary process, the outcome of competition among cell lineages—clones—that have escaped normal control through mutation. Chapter 6 describes the origins, nature, and consequences of cancer as an evolutionary process, unique in each patient, with a history that can be traced in the genome, and with major implications for treatment.

Natural selection can only optimize the parts of organisms that are not involved in conflicts, and reproduction is loaded with conflicts. Chapter 7 discusses them: conflicts between mother and offspring, between maternally and paternally derived genes in the offspring, and among siblings. It also discusses the evolution of menstruation and menopause and the connections between the evolution of placentas and susceptibility to metastatic cancer.

Cultures evolve much more rapidly than biology, putting biology in the position of trying to keep up with an accelerating process that it will never be able to catch. As a result our bodies are mismatched to our current environments. Chapter 8 discusses the many consequences, which include insights into obesity, cardiovascular disease, and autoimmune diseases.

A mind is not only a terrible thing to waste: it is a difficult thing to understand. Chapter 9 discusses evolutionary hypotheses about mental disease that give potentially helpful perspectives on addiction, anxiety, depression, obsessive-compulsive disorder, autism, and schizophrenia. These conditions all have multiple causes. The evolutionary contributions to those causes are still being sorted out; we report on the progress being made.

Despite the dramatic reductions in mortality rates that have resulted from clean water, vaccines, and antibiotics, humans continue to evolve, as do their pathogens and their cancers. Medical interventions have major consequences for evolution, and Chapter 10 explores them, in the process showing how medicine is creating some new problems while solving old ones.

The book ends in Chapter 11, which starts with a look at important questions raised by evolutionary insights that have not yet been answered. This section is for current and potential researchers, and anyone else who likes a mystery. This final chapter also discusses why we have not chosen to address some issues. It concludes with a comparison of classical with evolutionary medicine.

Suggestions for further reading

Inquiry into evolutionary answers for medical questions was reinvigorated by Paul Ewald with *Evolution of Infectious Disease* and Randy Nesse and George Williams with *Why We Get Sick: The New Science of Darwinian Medicine*. Their books raised many of the issues addressed here. Peter Gluckman, Alan Beedle, and Mark Hanson have written one text, *Principles of Evolutionary Medicine*, and Robert Perlman has written another, *Evolution and Medicine*. Daniel Lieberman has discussed the recent history of our body and the concept of mismatch, in depth and at length, in *The Story of the Human Body*. All those books invite your attention. In addition, there are two multi-author volumes, one, *Evolutionary Medicine and Health: New Perspectives*, edited by Wenda Trevathan, E. O. Smith, and James McKenna that is weighted towards anthropology, the other, *Evolution in Health and Disease*, edited by Stephen Stearns and Jacob Koella, that is weighted towards genetics and evolutionary epidemiology. There is much material for further exploration.

Thoughts for instructors

The job of the teacher is no longer primarily to transmit information: students acquire it efficiently from many sources. When we teach this material, we use class time for active discussion, and we do not allow students into the classroom if they have not done the readings or looked at online material. That material includes a set of short lectures that introduce the material in this book; they can be accessed at YouTube and at iTunesU—search for "Stearns Evolution Medicine."

We use the information in this book as the platform from which to launch students into the original scientific literature. We ask undergraduates to pick a topic, find the current state of knowledge in the recent literature, and write a critical paper on it. That gives them practice with a tool that can keep them learning for the rest of their lives. We ask medical students to select an appropriate entry in Wikipedia and improve it substantially. That sharpens a tool that many use. Both types of writing exercises give students agency, the feeling that they are taking control of their own lives. Transmitting that feeling is more important than transmitting information, for anyone who internalizes that feeling will be motivated to acquire all the information she needs.

Acknowledgments

We would like to thank our colleagues for their generosity in sharing ideas and images, in particular Peter Ellison, Peter Gluckman, Randy Nesse,

Charlie Nunn, and Cynthia Beall. Steve thanks all the students who have taken his courses in evolutionary medicine; they have been and continue to be a very stimulating bunch. The support we have received from Sinauer Associates has been superb; we thank in particular Andy Sinauer for his support of this project, Joanne Delphia for the book and cover design, Janice Holabird for paging the book, Stephanie Bonner for coordinating the editorial side of production and using her sharp eye to locate issues in the text, Kathleen Lafferty for editing the copy, and Sharon Hughes for making the index. They made a great team to work with.

Media and Supplements
to accompany *Evolutionary Medicine*

eBook

Evolutionary Medicine is available as an eBook, in several different formats, including VitalSource CourseSmart, Yuzu, and BryteWave. The eBook can be purchased as either a 180-day rental or a permanent (non-expiring) subscription. All major mobile devices are supported. For details on the eBook platforms offered, please visit www.sinauer.com/ebooks.

For the Instructor

Instructor's Resource Library

(Available to qualified adopters)
The Evolutionary Medicine Instructor's Resource Library provides instructors with a collection of visual resources to aid in preparing their course and presenting their lectures. The IRL includes the following:

- Textbook Figures & Tables: All textbook figures and tables in JPEG (both high- and low-resolution) and PowerPoint formats.

- Lecture Presentations: For each chapter of the textbook, the authors have prepared ready-to-use lecture presentations that include text reviewing the key facts and concepts from the chapter, along with selected figures and tables.

Introduction to Evolutionary Thinking

This chapter offers a brief overview of some key elements of evolutionary thought. We focus on the intellectual tools most frequently needed in evolutionary medicine. Our overview is not intended to be comprehensive.

Natural selection

Nature presents us with a puzzle: structures, organs, and processes whose complexity and precision suggests that they were designed, without a designer. They were produced by natural selection, a process in which no agent actively selects anything. It operates whenever anything correlates variation in reproductive success with heritable trait variation.

The four necessary conditions

There are four conditions under which natural selection on traits will occur:

1. There is variation in reproductive success.
2. There is variation in the trait of interest.
3. There is a nonzero correlation between the trait and reproductive success.
4. The state of the trait is heritable.

When in doubt, return to these basic conditions. They are useful criteria against which to judge claims of adaptation. Now let's unpack them.

Variation in reproductive success

To a first approximation, human variation in reproductive success is variation in completed family size, or children ever born. In bacteria and cancer cells, it is composed of variation in time and survival between cell divisions.

Natural selection is not about the survival of the fittest but about the increase in frequency of the reproductively successful. Reproductive success is a composite of survival and reproduction: the probability of surviving to reproduce multiplied by the number of offspring produced in that reproductive event. Selection acts directly through reproductive success but only indirectly through survival: survival is only important in evolution to the degree that it contributes to reproductive success. This insight is a key element of the evolution of aging, which results in susceptibility to degenerative disease.

Variation in reproductive success is universal. In any large group of people, there are some that are single children, some that have one brother or sister, and some that come from larger families. Although mortality rates have fallen and maximum family sizes have decreased in postindustrial societies, variation in completed family size remains substantial, and in some societies, such as Pitcairn Islanders or the Dogon of Mali, it remains strikingly large (Figure 1.1).

Variation in traits

Any trait that correlates with reproductive success will experience natural selection. Whether it responds to selection with change in succeeding generations will depend on whether it is heritable. The trait could be any part of the organism: eye color, height, the temperature dependence of enzymatic function, the structure of the ribosome, susceptibility to a disease, the efficiency with which a drug is metabolized, the tolerance of a pathogen. If the trait does not vary, there can be no competition among trait variants for improved reproductive performance and thus no natural selection: natural selection does not operate on things that do not vary. It is not currently shaping limb number in mammals because all mammals that manage to reproduce have two forelimbs and two hindlimbs (there are rare mutants with more and with fewer).

The many traits that do vary among individuals all potentially experience selection. One such trait, resistance to malaria, is determined in part by an individual's genotype at the sickle-cell locus: *SS* (sickle-cell homozygote), *SA* (sickle-cell heterozygote), or *AA* (normal homozygote). *SS* homozygotes suffer from complications of anemia that, if not treated, increase mortality. *AA* homozygotes have greater risk of dying of malaria. It is the *SA* heterozygotes that have the best chances of surviving malaria. The S variant is caused by a mutation in the sixth position of the β-globin chain of hemoglobin, a mutation that results in the replacement of hydrophilic

(A) Monogamous Pitcairn Islanders

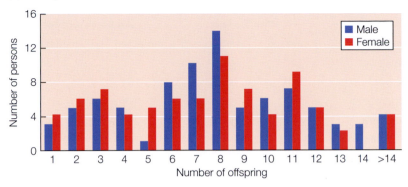

(B) Polygynous Dogon in Mali

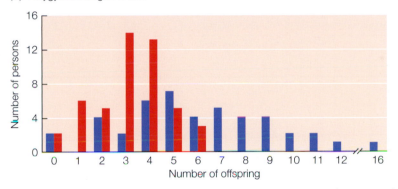

Figure 1.1 Variation in reproductive success among men and women who have completed their families in two preindustrial societies. (After G. R. Brown, Laland, and Mulder 2009.)

glutamine by hydrophobic valine, altering both the shape of the protein and the red blood cells that carry it.

Across sub-Saharan Africa, populations vary in the proportion of each genotype present and infected (Table 1.1). Natural selection driven by malaria is favoring the maintenance of the sickle-cell polymorphism despite the costs imposed on sickle-cell homozygotes. As shown in the table, sickle-cell carriers in sub-Saharan Africa were 6.2% less likely to be infected by malaria than were normal homozygotes.

The correlation of traits with reproductive success

Even if individuals vary in reproductive success and have a trait that varies, it does not yet mean that the trait experiences selection. Those conditions are necessary, but they are not sufficient. Also needed is a correlation between the state of the trait and reproductive success. For example, individuals who better resist disease may reproduce more successfully than those who do not resist as well.

TABLE 1.1 Variation in sickle-cell genotypes across sub-Saharan Africa

Population	Region	With S allele		AA homozygotes	
		Infected	Uninfected	Infected	Uninfected
1	Uganda	12	31	113	134
2	Kenya	131	110	154	87
3	Uganda	73	118	494	515
4	S. Ghana	42	131	270	572
5	N. Ghana	11	4	165	12
6	Nigeria	162	51	680	210
7	N. Ghana	13	6	109	18
8	Nigeria	51	40	245	97
9	Tanganyika	77	59	272	135
10	S. Ghana	34	89	176	417
Total		606	639	2678	2197
Percentage		48.7%	51.3%	54.9%	45.1%

Source: Allison 1964, Table 1, p. 138.

It is such correlations between variation in reproductive success and variation in traits that drive natural selection. What in general causes them? What thus causes the subsequent appearance of design? There is no single general cause of selection. Selection occurs for a variety of reasons that encompass anything yielding a correlation between a trait and reproductive success; these are countless circumstances in the biology of organisms and, in humans, also in their culture, one element of which is medicine. Medical practice and public health measures have become significant sources of natural selection, which continues to act in contemporary human populations.

For example, women born in Framingham, Massachusetts, between 1890 and 1960 varied both in reproductive success and in several traits of medical interest. Analysis of the correlations between their traits and their reproductive success revealed that selection was acting to decrease total cholesterol, blood pressure, height, and age at first birth and to increase weight and age at menopause (Byars et al. 2010).

Variation must have a genetic basis

Selection may act on traits, but it will not produce a response unless those traits vary genetically. The state of the trait must be heritable. This critical genetic condition allows what has worked in the past to be "remembered," permits improvements to accumulate, and generates order out of disorder.

However, inheritance must not be absolutely perfect, for if it were, genetic variation would not be originating through mutation. There must be some mutation to create the genetic variation that is the fuel on which the motor of natural selection runs, but not so much mutation that it destroys the memory of what has worked in the past. Selection adjusts the mutation rate to an intermediate level that balances these two considerations. At that level, the mutation rate remains significant enough to enable the evolution of populations of organisms and clones of cancer cells (whose mutation rate is often much higher than would be optimal for an organism).

Evolutionary geneticists measure genetic variation in phenotypic traits by analyzing the correlations among relatives and expressing the results on a scale that runs from zero to one. Those values are called the heritabilities of the traits; the symbol for heritability is h^2. Heritability has a technical meaning that differs from its meaning in everyday language. A heritability of 1.0 means that all offspring have exactly the same value of a trait as the average of the two parents. A heritability of 0.0 means that relatives do not resemble each other any more than would two unrelated people chosen at random. Heritability measures the proportion of total phenotypic variation in the population that can be accounted for by the effects of many genes acting additively, unconstrained by interactions with other genes. Other than genetic effects, the major source of trait variation among individuals is the environment in which they have been raised.

The heritabilities of several human traits have been measured many times in large populations (Stearns et al. 2010). For age at first birth, $h^2 =$ 0.11 (mean of 5 estimates, $n \approx$ 12,000); for blood traits including cholesterol, $h^2 = 0.36$ (mean of 66 estimates, $n \approx$ 15,000); for age at menopause, $h^2 =$ 0.59 (mean of 24 estimates, $n \approx$ 14,000); and for height, $h^2 = 0.75$ (mean of 115 estimates, $n \approx$ 130,000). These traits all have enough genetic variation to respond to selection. For a given correlation with reproductive success, height will respond more rapidly than age at first birth because the response to selection is determined by the product of heritability with selection intensity, the strength of the correlation with reproductive success.

Thus, humans vary in reproductive success, many human traits vary among individuals, there is often a nonzero correlation between some of those traits and reproductive success, and most traits have heritabilities large enough to permit a response to selection. In short, humans satisfy all the conditions for natural selection. They continue to evolve. So do their pathogens, only much more rapidly.

An example: Antibiotic resistance

Bacteria rapidly, repeatedly, and reliably evolve resistance to antibiotics. Many of the genes that confer resistance evolved long ago in the context of coevolutionary struggles among bacteria and between bacteria and fungi; they do not need to arise through mutation. Large genetic libraries of information on how to resist antibiotics thus exist in nature. Many of the genes involved can be horizontally transmitted, allowing evolving bacterial populations to acquire and accumulate resistance to multiple antibiotics rapidly. Some of them, such as *Staphylococcus aureus*, have evolved strains that cannot be treated by any existing antibiotic: they have become "superbugs."

The history of resistance evolution in *S. aureus* has gone as follows:

- 1943   Penicillin commercially available
- 1947   First resistance reported
- 1960s Switch to methicillin
- 1980s Methicillin resistance rising
- 1990s 35% of isolates resistant to methicillin
- 1990s Switch to vancomycin
- 1996   Vancomycin resistance reported
- 2000   Linezolid approved by the U.S. Food and Drug Administration
- 2002   Linezolid resistance reported

This sad history exemplifies the coevolutionary treadmill of drug resistance. A drug is released; its success leads to widespread use; bacteria evolve resistance to it, rendering it useless; another drug is introduced; and the cycle repeats. It happens quickly and for good reasons. Bacterial populations are huge; they have access through horizontal gene transfer to pre-evolved libraries of genetic information on resistance; they generate vast numbers of genetic variants through mutation; and their generation times are short, measured in hours or days. By creating huge differences in the reproductive success of resistant and nonresistant clones, antibiotic use generates very strong selection. Resistance is the trait that varies, that variation has a genetic basis, and selection is strong. The process is rapid and efficient. Within a few years, resistance has evolved against every new antibiotic that has been introduced. We are starting to move into a post-antibiotic era in which treatment of bacterial infections is more difficult and surgery is riskier because postsurgical infections are becoming more difficult to treat than previously.

Other cases of medical importance

Antibiotic resistance is first in order of importance on the growing list of examples of natural selection in clinical medicine and public health. The list also includes the following:

- The evolution of metastasis in cancer
- The evolution of resistance to cancer chemotherapy
- The coevolution of pathogens with host defenses, leading to evasion and manipulation
- The evolution of pathogen virulence in response to changes in transmission and to vaccination campaigns
- The effects of medical practice on human, pathogen, and cancer evolution

Natural selection is thus both a cause of medical conditions and a consequence of medical practice.

■ SUMMARY

Natural selection occurs whenever variation in a trait is correlated with variation in reproductive success. Traits experiencing selection only respond to it if their variation is heritable, and that response will only be significant if the events that generate it occur frequently and consistently. When that happens, selection efficiently creates order out of disorder, producing the precision and complexity of design that we call *adaptation*. No agent actively selects. The process emerges, unsupervised, from any circumstances that correlate heritable variation in traits with reproductive success.

Neutral evolution

Evolution occurs whenever genes change in frequency in populations. They can do so either because they are selected, as discussed above, or because random processes affect them. Random processes dominate when genetic or phenotypic variants do not differ in their effect on reproductive success, that is, when their variation is *neutral* with respect to fitness. Neutrality arises for different reasons at the genetic and at the phenotypic levels. At the genetic level, many genotypes may produce the same phenotype because the genetic code is redundant, because some DNA is not expressed, because some amino acid substitutions produce no change in the function of a protein, and because development is canalized. At the phenotypic level, many phenotypes can have the same fitness because their trait variation makes no difference to reproductive success or because the effect on fitness of a change in one trait compensates precisely for that of a change in another trait. Let's unpack each of those conditions.

The genetic code is redundant

The genetic code is redundant because it is a triplet code in which each position in the triplet can be occupied by one of four nucleotides, generating (4 × 4 × 4) = 64 possible codons, but those 64 codons are only being used for 23 meanings: to signify 20 amino acids and 3 stop codons. In consequence, several triplet nucleotides code for each amino acid. For example, leucine is coded for by 6 triplets: UUA, UUG, CUU, CUC, CUA, and CUG (U = urine, A = adenine, G = guanine, and C = cytosine). A mutation from A to G in the third position, changing UUA to UUG, would not result in any change in the amino acid expressed, which would remain leucine. Such mutations are called synonymous substitutions; their evolution is neutral, driven by random processes (unless they affect noncoding properties of DNA or RNA). It is primarily the nonsynonymous substitutions that have known fitness consequences.

That random processes drive genetic change in part of the genome—particularly in the third position in codons—is extremely useful for the reconstruction of relationships and the inference of history. It means that the amount of change between two lineages is proportional to the time that has elapsed since they shared a common ancestor.

Some DNA is not expressed

Eukaryotic genomes contain a considerable amount of DNA that is not expressed either in proteins or in regulatory RNA. Some of the unexpressed DNA is a legacy of duplicated genes that have not acquired a function—pseudogenes. Some of it resulted from viruses that inserted themselves into the genome and could not re-activate. For several reasons, much of the eukaryotic genome is neutral because it is not transcribed.

Some amino acid substitutions do not change protein function

Neutral change can also occur at the level of amino acid substitutions in proteins if those substitutions do not produce significant, functional change in the shape or charge of the protein. One highly functional molecule shared by all vertebrates is hemoglobin, which transports oxygen in blood. It consists of two α-globins and two β-globins attached to iron-containing heme groups. The amino acids in the α-globins differ among vertebrate species, and we can estimate the time since those species shared common ancestors from fossils. The relationship between the number of amino acid differences and the time available for change is linear (Figure 1.2), indicating that the rate of amino acid substitution is uniform in time.

   Such a constant substitution rate suggests a random process. The amino acid substitutions that make a significant functional difference have for the most part been detrimental, have not accumulated, and are therefore not represented in Figure 1.2. This evidence indicates that changes in a portion of the hemoglobin molecule, but not all of it, make no difference to function. It also suggests that neutral change can be used as a molecular clock to estimate time to last common ancestor.

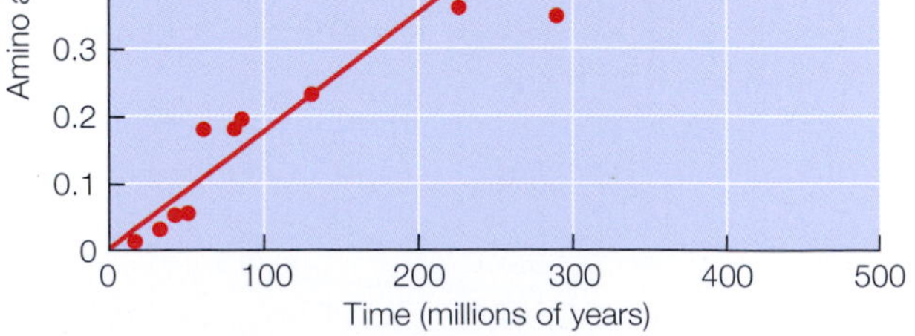

Figure 1.2   Differences among vertebrate species in the amino acids found in α-globins have accumulated at a constant rate since those species shared ancestors, indicating that those differences are close to neutral. Time is estimated from fossil samples. (After Kimura 1983.)

Canalization buffers phenotypes against genetic and environmental changes

Another reason phenotypes may not change when genes vary is canalization. Canalization describes developmental processes that limit phenotypic variation. These processes buffer the phenotype against genetic and environmental perturbations. Genes whose effect on the phenotype has been reduced or eliminated by canalizing mechanisms are freer to accumulate neutral variation through mutations that are not scrutinized by selection than are genes whose effect is directly expressed. Canalized traits in tetrapods include four limbs, two eyes, one mouth, and five digits. Genetic variation affects many other things about those traits, but it only rarely affects their number, and when it does, the consequences are disastrous.

Some phenotypic variation has no effect on reproductive success

It is also possible for phenotypic variation to have little or no effect on reproductive success. It could be a type of variation that cannot be detected by mates, predators, pathogens, or prey, or it might be a type of variation that lies within limits that permit normal function. For example, it probably makes little difference to fitness whether a person is 165.1 or 165.2 centimeters tall or has a few hairs more or less on the back of the neck.

Trade-offs cause fitness compensation among traits

Reproductive success is the composite result of contributions from multiple traits. The more precisely those traits compensate for each other's effects on fitness through *trade-offs*, the closer the trait variation is to neutrality. A trade-off occurs whenever two or more traits are linked in such a way that an increase in fitness caused by an evolutionary change in one trait is accompanied by a decrease in fitness caused by an evolutionary trait in another trait. (We discuss trade-offs at greater length in Chapter 2.)

For example, the same lifetime reproductive success can be achieved by many combinations of number of offspring trading off with offspring survival (more offspring, worse individual survival). Other examples are success in fighting for mates trading off with resistance to disease (better mating success, worse resistance) and better resistance to normal infection trading off with greater costs of inflammation.

The costs and benefits of the inflammatory response were starkly revealed in the 1918 influenza pandemic, in which the mortality rate was highest in young adults. Their strong and normally advantageous responses caused their lung alveoli to flood with inflammatory exudate, leading many to die of secondary bacterial pneumonia.

■ SUMMARY

Neutral evolution occurs because there are disconnects between DNA sequence variation and variation in fitness that occur at a series of levels in gene expression and organismal development. Those disconnects include the redundancy of the genetic code, amino acid substitutions that do not change protein function, developmental canalization, variation in traits that has no effect on reproductive success, and compensating variation in traits involved in trade-offs.

Mechanisms causing random change

Randomness enters the evolutionary process in at least four important ways: the sense in which mutations are random, the effects of population sampling (founder effects and genetic bottlenecks), the Mendelian lottery (meiosis is like flipping a fair coin), and variation in reproductive success that is not affected by the focal gene or trait for any of the reasons given above (which happens in populations of all sizes, not only small ones). These last two mechanisms—the fairness of meiosis and neutral variation in reproductive success—drive genetic drift.

The sense in which mutations are random

Mutations are *not* random in several ways: they occur more frequently at some sites in the genome than others; the overall mutation rate in some pathogenic bacteria increases in response to signals of stress, including attack by the immune system; mutations causing transitions among nucleotides (purine to purine, pyrimidine to pyrimidine) are more common than transversions (purine to pyrimidine, pyrimidine to purine); and mutations cannot produce random changes in phenotype space, by which we mean, for example, that the phenotypic changes that mutations can produce in a fly are different from those that mutations can produce in a mouse (mutational variation is constrained by body plans). Those are all important ways in which mutations are not random.

There is, however, an extremely important sense in which mutations are random: they are random with respect to their effect on reproductive success, on fitness, on the needs of the organism. Mutations do not anticipate what would be useful; rather, they create variation that is then edited by natural selection.

Effects of small populations

If a small sample is taken from a large population, it can only contain an amount of genetic variation smaller than that contained in the larger population. A *founder effect* occurs if that small sample is generated by emigration to a new site. For example, the unusually high rates of Tay-Sachs disease in Quebecois, of porphyria in Afrikaaners, and of diabetes in Pitcairn Islanders were caused by the presence of genes for those diseases, which are normally rare, in the small founding populations.

A *genetic bottleneck* occurs if the small sample is caused by a population crash to a very small size. Only a small amount of genetic variation makes it through a bottleneck because only a few individuals survive to carry it. For example, the homozygosity of immune genes in cheetahs, confirmed by the ease of reciprocal skin transplants, was probably caused by a genetic bottleneck.

Populations that remain consistently small experience a third important effect. Traits under significant selection in large populations, where large numbers of events occur consistently, encounter neutral evolution in small populations, where the small sample of events does not permit a consistent correlation between reproductive success and traits to emerge. Selection can discriminate smaller differences in functional traits in larger populations, differences that would be neutral in smaller populations. In other words, a mixture of natural selection and random processes drives the evolution of all genes and traits. Functionality and neutrality are dependent on situation: sometimes selection dominates, sometimes randomness.

Meiosis is a fair coin

Gregor Mendel's law of segregation states that if there are two alleles present at a locus in a heterozygotic diploid organism, the probability that one of them will get into a given gamete is 50%. The precision of that probability, perhaps most easily seen in the chance of having a son or a daughter, is impressive. (It probably evolved because the majority of genes mobilized defenses against selfish minorities that sought to distort transmission for their benefit at the expense of majority interests.)

In the long run, the expected reproductive success of a diploid sexual individual is just two surviving offspring, a conclusion derived from the observations that populations remain within limits and that any other value would, in the long term, lead either to extinction or unsustainable population explosion. That means that if you are a gene sitting on a chromosome in a diploid sexual individual, you have, as a long-term average, just two chances, two flips of a fair coin, to get into the next generation. Sex is a risky business for genes; in every generation, they stand a fair chance of not being transmitted into an individual that survives and reproduces.

Genes drift when they land at random in families of different sizes

When genetic variation has no effect on reproductive success, the neutral alleles land at random in families of different sizes, encountering random variation in reproductive success: that is the meaning of zero correlation with reproductive success. Across a sequence of generations, it causes their frequencies to increase and decrease at random, like the Brownian movements of minute particles buffeted by thermal molecular motion in a liquid. This process—called *genetic drift*—happens in populations of all sizes, large as well as small.

Genetic drift is driven by the variation in reproductive success that neutral alleles encounter when they land in families of different sizes at random. The usual fate of a drifting, neutral mutation is to disappear from the population because it is initially at very low frequency, and its probability of fixation—of increasing in frequency until it occurs in every individual in the population—is equal to its frequency. A few neutral mutations manage to run the gauntlet of repeated random family sizes and end up being fixed.

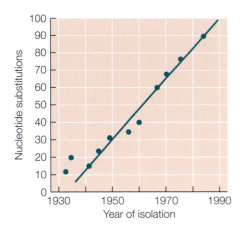

Figure 1.3   The regular fixation of neutral mutations in influenza samples produces a molecular clock. (After Gojobori, Moriyama, and Kimura 1990.)

There are fewer mutations in smaller populations, but drift fixes mutations faster in smaller populations. In fact, it does so at a rate that precisely compensates for population size. The result is that neutral mutations get fixed at a constant average rate that is independent of population size. We do not know *which* mutation will be fixed in any given period of time, but we do know *how many* will be fixed on average. In that sense, the molecular clock that ticks every time a neutral mutation is fixed is like an atomic clock driven by radioactive decay: in any given period, we do not know *which* atom will decay, but we do know *how many* on average will decay. In both cases, regularity emerges from a very large number of independent random events. The haploid human genome consists of about 3×10^9 base pairs, and one mole of uranium contains about 6×10^{23} atoms.

The molecular clock seen in influenza samples

The regular ticking of the molecular clock has been clearly seen in samples of the influenza virus isolated and frozen since 1930. Pathogens causing epidemics experience huge fluctuations in population size. Despite those fluctuations, the rate of accumulations was so steady and linear that sequencing an unknown sample would give a good estimate of the date it was isolated (Figure 1.3). The effects of fluctuating population sizes on the numbers of new mutations introduced per generation exactly compensated for the slower rate of fixation of neutral mutations in larger populations.

■ SUMMARY

For several reasons, variation in DNA and variation in traits may make no difference to reproductive success. Such genes and traits are neutral with respect to fitness. Two mechanisms determine the fate of neutral mutations: meiosis and random variation in family size. Meiosis is like a fair coin: the long-term average probability of getting into a gamete that helps produce an organism that survives and reproduces is 50%. When mutations are neutral, they are transmitted at random through families of different sizes. Those two effects combine to produce genetic drift, which causes the frequency of neutral mutations to fluctuate at random and fixes neutral alleles at a regular rate independent of population size. The fixation of neutral alleles is like radioactive decay. In neither case do we know which mutation will be fixed or which atom will decay, but because there are so many of them, we do know how many events will happen on average in a unit of time. That regularity of fixation of neutral alleles allows us to estimate times to last common ancestors by measuring how many differences have accumulated in the neutral portions of DNA sequences.

Mismatch

Mismatch occurs when organisms that were well adapted to one set of environmental circumstances cannot evolve rapidly enough to adapt to a new set of circumstances. Of the big ideas in evolutionary medicine, mismatch is one often invoked but frequently misused. For example, many people concerned with nutrition and exercise as components of a healthful lifestyle like to imagine an ancient environment, generically called the environment of evolutionary adaptation or "EEA," to which we were once well adapted and deviations from which are the cause of the "diseases of civilization" that we now experience: obesity, cardiovascular disease, diabetes, autoimmune diseases, and allergies, not to mention anxiety, depression, psychosis, and other mental disorders.

Although there is certainly some truth to this view—mismatch is real and has important consequences (Lieberman 2013)—it is also a view that has been applied too often without rigor. Partisans of particular hypotheses have often invoked environments of evolutionary adaptation that would support their ideas without being able to present any evidence that those environments actually existed at high enough frequency to play important roles in selection. In a few important cases—lactose tolerance, autoimmune diseases, allergies—we do have such evidence and are able to make a strong case for mismatch (see Chapter 8). For others, such as mental illness, the approach is interesting but speculative.

Mismatch in time and space

Mismatch in time occurs when the environment changes faster than the population can adapt. The most dramatic recent changes in the human environment have been caused by cultural innovations: agriculture, urbanization, and hygiene. The agricultural revolution produced a dramatic shift in the human environment. It enabled the growth of cities, city-states, and the eventual emergence of nations; it marked the transition from clans and tribes to urban collections of unrelated strangers; it caused a huge change in diet; and through exposure to the pathogens of domesticated animals and the significantly increased possibilities for horizontal transmission of pathogens and parasites, it enabled the emergence of many diseases. The more recent industrial and postindustrial revolutions, bringing with them clean water supplies, vaccination, antibiotics, and other medical and public health improvements, have caused significant drops in mortality rates and then, after a period during which populations exploded, drops in birth rates. In most cases, those cultural changes have happened faster than biology could adapt.

Mismatch in space occurs when organisms move from environments in which they are well adapted to environments in which they are not. In humans, it happens through the reciprocal processes of emigration and immigration. A good example of mismatch in immigrants is the prevalence of sickle-cell disease among African Americans. The sickle-cell gene is advantageous where malaria is a significant cause of mortality, but where malaria is absent, it is a serious genetic disease without compensating benefits.

We are therefore mismatched to things encountered rarely or never before. We are also mismatched to things that happen in other environments but not this one.

Traits are adapted to things that happen frequently

Things that happen frequently can lead to adaptations, but rare events seldom do so. We have to think about frequency from the point of view of genes, which are much more frequently in the young than they are in the old and which come much more frequently out of supportive environments and healthy bodies than they do out of threatening environments and sick bodies. We are adapted to the circumstances that have frequently led to success in the past.

Another important point about frequency is that in a hierarchical selection system, things that happen very frequently but with small effects at lower levels in the hierarchy can accumulate to have a significant effect on fitness at a higher level. For example, protein synthesis happens many times per second in virtually every cell in the body, and our bodies contain about 10 trillion cells. A difference in ribosome structure that improves protein synthesis by, say, 0.0000001%, which looks trivial when we look at a single cell for the brief moment that it takes to synthesize a protein, can accumulate to make a much bigger selective difference at the level of the lifespan of a whole organism.

■ SUMMARY

Mismatches between the evolved state of the organism and the environment it is currently encountering occur for three reasons: the environment changes too rapidly for biology to keep pace, organisms move from the habitat in which they evolved into one to which they have not yet adjusted, and the environment is novel. All three effects occur in humans. Changes in time have been driven primarily by cultural evolution, in particular agriculture, urbanization, industrialization, and modern medicine and public health. Changes in space are experienced by immigrants. Such experiences elicit a broad range of significant medical issues. The plausibility of the logic and the potential resulting effects, however, have led some to invoke mismatch in cases in which it has not yet been demonstrated.

Adaptation

Those first encountering the explanatory power of natural selection sometimes get carried away. They have a tendency to see adaptation everywhere and make selection the central character in "just-so" stories. Claims of adaptation should not be accepted unexamined; rather, they should be tested against alternatives. The states of patients and pathogens that medical science encounters are not all adaptive. They may be by-products of trade-offs, results of random processes, maladapted mismatches, or the results of constraints that selection could not overcome.

Evolutionary biologists have responded to this issue by developing criteria for recognizing adaptation, including observing natural selection in action; perturbing the system and observing the fitness consequences; seeing whether the trait is only produced when it serves a function; asking whether its complexity and precision suggest a long history of many selective events; and, when the system can be modeled, testing the observed state for its ability to resist invasion by competing alternatives. Claims of adaptation that cannot be tested against any of those criteria should be distrusted. Let's now take a closer look at the criteria.

Observing natural selection

If one can observe heritable changes in a trait that result from a demonstrated correlation of the state of the trait with reproductive success, there is no doubt that the *change* in the trait is an adaptation. This criterion is difficult to fulfill because it requires measuring traits, heritabilities, and lifetime reproductive successes over enough generations for a genetic response to selection to occur. That is why it has seldom been fulfilled in humans. These properties are easier to observe in experimental evolution in microorganisms, such as viruses and bacteria, and other short-lived model organisms, such as flies and nematodes. The idea is simple: if you can actually watch the process and see the response in a controlled experiment, you can be sure that the response was produced by the process.

Perturbing the trait

Both a predictive model and an experiment are required to assess adaptation by perturbing a trait. The model is used to predict the optimal state of the trait, and the experiment tests that prediction by perturbing the trait from the optimal state. If the experiment demonstrates that the fitness of the organisms with the perturbed trait is lower than that of those with the predicted trait, the predicted state is adaptive. Daan, Kijkstra, and Tinbergen (1990) did such an experiment on clutch size in kestrels. The biologists removed eggs from some clutches, added eggs to other clutches, and did similar manipulations, removing and then returning eggs to the same clutches, in the controls. They then followed the survival and reproduction of both the parents and the offspring. The results are shown in Table 1.2.

To measure the outcomes, Daan and his colleagues calculated the reproductive value V of the kestrel fledglings, the broods, the clutches, and the parents. Reproductive value is the number of offspring expected from that point in life onward. As shown in Table 1.2, enlarged clutches produced more offspring than reduced clutches, but the parents of enlarged clutches suffered the cost of having to care for more offspring: they were less likely

TABLE 1.2 Impact of brood size on fitness in kestrels

	Reduced	Control	Enlarged
Brood:			
Number of broods	28	54	20
Mean number fledged	2.60	3.95	5.84
Clutch reproductive value V_c	3.16	5.25	6.99
Parents:			
Number of males and females	49	85	35
Parent local survival	0.653	0.588	0.429
Parent reproductive value V_p	10.14	9.14	6.69
Total reproductive value ($V_c + V_p$)	13.30	14.39	13.68

Source: Daan, Dijkstra, and Tinbergen 1990, Table 3, p. 97

to survive to the next year. Over the course of their lifetimes, those costs more than compensated for the additional offspring that they got in the one year in which their clutches were manipulated. As a result, the birds with the control broods got, on average, about one more offspring per lifetime than did those with the reduced broods and about 0.7 more offspring per lifetime than those with the enlarged broods. The biologists found that kestrels did lay the number of eggs that maximized their reproductive success; thus, the clutch sizes were adaptive.

Producing the trait only when it serves a function

If a trait responds to a specific environmental signal with a change that has a clear functional relationship to the signal and improves reproductive success but otherwise does not respond because the change is costly, we have reasonable evidence that the change in the trait is an adaptation (Williams 1966). Some convincing cases of such changes are called induced responses. The water flea *Daphnia* produces offspring that grow spines and helmets when the water flea detects dissolved molecules associated with predators whose efficacy in eating *Daphnia* is reduced by spines and helmets. Those defenses have a reproductive cost: water fleas in defensive mode produce fewer eggs than those in nondefensive mode. Barnacles respond in a similar fashion when they sense the presence of snail predators: they grow bent shells that make them resistant to predation, and they pay a cost for their resistance in reduced fecundity.

A case with medical implications involves the digenetic trematode worm *Schistosoma mansoni*, which causes schistosomiasis. This worm has a complex life cycle in which it alternates between infecting aquatic snails and infecting humans or other vertebrate hosts. When it infects snails, it castrates them, which has the advantage, for the worm, of causing its host to shift investment from reproduction to growth, thus producing more tissue for worms to feed on and, eventually, more offspring. Snails exposed

to water in which *Schistosoma* had been held but was no longer present shifted their reproduction earlier in life; they had sensed danger and
responded to it by reproducing before they would be castrated. Such
induced responses can be convincingly called adaptations.

The design criterion

Often, we are not able to observe natural selection in action, we cannot
perturb the trait and observe the consequences, and the trait does not
change in response to environmental signals. The state of the trait might
still have been produced by natural selection, which is more likely if the trait
is precise and complex, conforming to a priori design principles. In other
words, if the trait resembles something that an engineer might design,
we can entertain the hypothesis that it is an adaptation. Such is the case
with any complex organ that performs a difficult function efficiently. For
example, in its dark-adapted state, the vertebrate eye is capable of detecting the arrival of photons produced by a single match at a distance of 10
kilometers.

To test the hypothesis that a complex, efficient, and precise trait is an
adaptation, we can try to answer these questions: Have experiments been
done to support the claimed function? Has the performance of the trait in
fulfilling that function been compared with alternatives? Do phylogenetic
analyses suggest that the state claimed to be an adaptation is repeatedly
associated with the kind of natural selection needed to produce it? Have
traits that started from different ancestral states converged on this state
every time they encountered similar environments? Could the trait have
been selected as a by-product of selection on other traits? Have we, in fact,
defined the trait correctly, or have we inappropriately abstracted a piece of
an organism from the larger whole in which it is embedded? That is a for-
midable series of questions. It is not often that they can all be convincingly
answered. When they cannot be answered, it is better to remain agnostic
than to insist, without adequate justification, that a trait is an adaptation.

Resisting invasion

In some cases, alternative states can be convincingly modeled. The claim
of adaptation can then be tested by allowing the modeled population to
be repeatedly invaded by possible alternative states of the trait as might
be generated by mutations. If the candidate adaptation can resist invasion
by competitors, we can say that it is plausibly an adaptation in the sense
that it is evolutionarily stable.

The criterion of evolutionary stability is useful, but it does not cover all
logical possibilities. If A beats B beats C beats A, the alternative states cycle
endlessly. It can also happen that the state that would resist all invaders
cannot be reached from any realistic starting point (e.g., a novel mutation
at initial low frequency). Not infrequently, another outcome occurs when
negative frequency dependence selects against the common and for the
rare, resulting in a mixture of phenotypic states and yielding an adapted
set no member of which is optimal.

■ SUMMARY

Observing a response to natural selection in action is the most convincing support for a claim of adaption. Perturbation experiments and induced responses are also convincing. The design criterion is plausible if some difficult questions can be answered, and resistance to invasion is a strong criterion if it can be tested against real alternatives. The bottom line is that claims of adaptation are problematic if they have not been tested. You should be skeptical of untested claims of adaptation and particularly skeptical of those that cannot be tested.

Styles of thought

To understand a new type of argument, it helps to know what implicit assumptions are being made, what different shades of meaning are being assigned to words, and what issues are hanging unexpressed in the background; in brief, it helps to know where people are coming from. At least three styles of thought are used when approaching issues in evolutionary medicine: typological thinking, population thinking, and tree thinking. We make them explicit here to help head off confusion later on.

Typological thinking

When thinking typologically, one has implicitly assumed that the most important thing to know about something is the category that it represents, the "normal" condition. This approach derives from Plato's theory of Ideals and Aristotle's theory of Types. The Type is thought to embody the essence of the thing; scientific examples include The Hydrogen Atom, The Vertebrate Kidney, and The Eukaryotic Ribosome. As the saying goes, if you've seen one, you've seen them all.

Typological thinking is useful whenever the average properties of a thing are much more important than its variation from the norm. That is often, if not always, the case in physics, chemistry, and molecular and cell biology. It is a powerful simplification that makes life easier, but because it does not always work, it is important to know when one can and cannot get away with it. It will mislead you when thinking about evolution, adaptive immunity, cancer, and any other process where variation within populations is an important driver.

Population thinking

In this mental mode, one assumes that the most important thing to know about something is the variation in the population from which it is drawn. It was Darwin's ability to think this way that led to his discovery of natural selection, which, among other things, selects the immune cells that produce antibodies specific to particular pathogens. Population thinking is also essential when dealing with the relative risk of heart disease stratified by age, sex, ethnicity, and lifestyle or with the risk of bad reactions to drug treatments as a function of genotype. It emphasizes variation in the population and the processes that change it, with individuals seen as samples whose traits can be estimated from population frequencies and sometimes determined by specific tests.

Population thinking is useful whenever it is important to know whether, and how fast, something will evolve. In medicine, that includes antibiotic resistance, pathogen virulence, cancer malignancy, and anything related to the microbiota. Population thinking forces us to consider frequency distributions, patterns of variation in space and time, and probabilities rather than certainties. It leads us always to ask how the other variants in the system are going to react when one of them changes.

In the case of antibiotic resistance, it is critical to know whether variation for resistance is present in the patient before treatment begins because in that case, antibiotic use eliminates the competition and favors the most resistant strains, which increase rapidly in frequency under treatment. This situation contrasts with the case in which no resistant strains are present before treatment starts and the effect of treatment in reducing the populations of the pathogens is to decrease the probability that a mutation for resistance will arise. To see which was the case in two humans who had not taken antibiotics for at least a year, Sommer, Dantas, and Church (2009) tested the ability of 572 genetically identified isolates from the gut microbiota to grow aerobically in the presence of 13 antibiotics administered at concentrations that prevent the growth of wild-type *E. coli* (Figure 1.4).

These results make clear that individuals who have not recently taken antibiotics have in their gut microbiota many strains of bacteria that carry genes for resistance against all the commonly used antibiotics (Figure 1.5). These bacteria are normally commensal, not pathogenic, but they could become pathogenic if they escaped the gut and colonized other tissues.

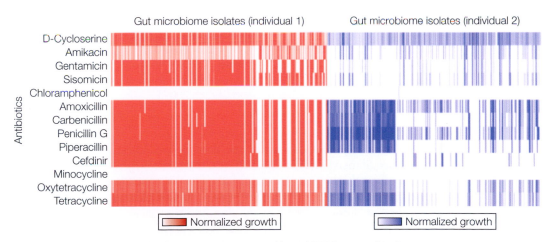

Figure 1.4   A heat map displaying resistance profiles of 572 bacterial isolates obtained from the gut microbiome of two individuals. White indicates no growth. (After Sommer, Dantas, and Church 2009.)

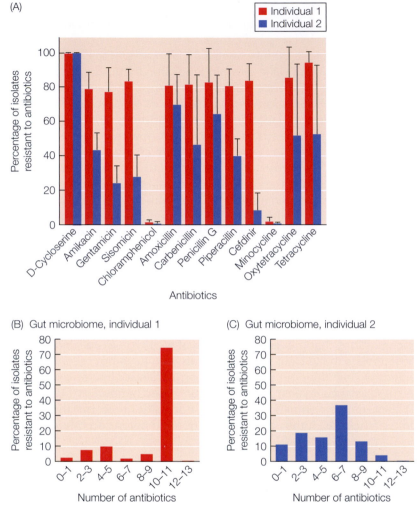

Figure 1.5   (A) The percentage of isolates resistant to each antibiotic. (B, C) The percentage of isolates resistant to the indicated numbers of antibiotics in the two individuals. (After Sommer, Dantas, and Church 2009.)

Giving infected individuals antibiotic therapy would rapidly select for the dominance of resistant bacteria and would lead rapidly to treatment failure.

Population thinking is also essential to understanding cancer as an evolutionary process (Figure 1.6). Every cancer arises as a population of clonal lineages that differ in their history of mutations. Some of those mutations (drivers) make functional differences to clone performance; others simply accumulate and are taken along for the ride, serving as markers without affecting outcomes. Selection allows some clones to expand; others go extinct or into dormancy. Different tissues constitute ecosystems to which

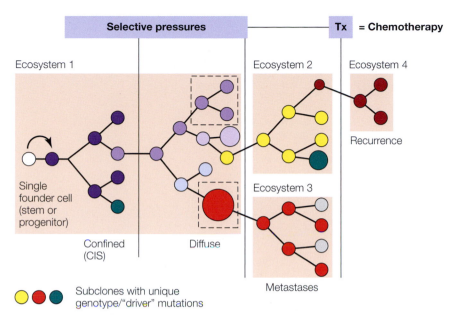

Figure 1.6 Every cancer is an evolutionary process driven by mutations inherited within competing clones that form a population of actors. Their differential performance in colonizing tissue shapes metastasis and their differential ability to resist chemotherapy (Tx) explains treatment failure and recurrence. (After Greaves and Maley 2012.)

different clones become adapted, and their efficiency of adaptation determines the progression of metastases. Chemotherapy selects for resistant clones, and their survival explains treatment failure and recurrence.

Tree thinking

In this mode of thought, the most important thing to know about something is its position in a phylogenetic tree, a tree that expresses the relationships among species or clones and thus traces a history with a record of the origins of similarities and differences. Because species and clones are not independent of each other but are connected by relationships, explanations are given in terms of how similarities are due to shared ancestry and how differences arise by divergence from the last shared ancestor.

Tree thinking is useful whenever insights can be drawn by comparisons among species or clones, that is, whenever history is an important component of the explanatory mix. In evolutionary medicine, such issues include the association of placental invasiveness with metastatic cancer, the association of upright posture with problems giving birth, understanding why elephants and whales do not all die of colon cancer as teenagers, why African green monkeys are frequently infected with simian immunodeficiency virus but do not get a disease from it comparable to AIDS, and why heart

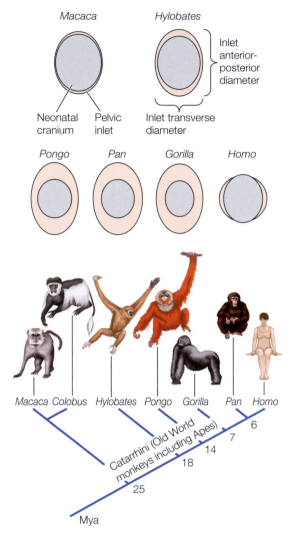

Figure 1.7 Phylogenetic trees guide us to the appropriate comparisons, in this case, between the birth canals of chimpanzees (*Pan*) and humans. (After Rosenberg and Trevathan 2002.)

disease presents very differently in chimpanzees than in humans. Knowing where on the tree of life a condition is present or absent immediately gives us clues to its correlates and possible causes.

For example, consider the relationship between the size of the neonatal head and the maternal pelvic inlet in six primate species (Figure 1.7). Looking at the phylogenetic tree immediately tells us which is the key comparison: it is between chimpanzees (*Pan*) and humans, not between gibbons (*Hylobates*) or macaques (*Macaca* and *Colobus*) and humans. That comparison tells us that since the time of our shared ancestor with chimpanzees, our neonatal head has expanded to more than fill the pelvic inlet. The difficulties that humans encounter in giving birth are therefore a recent condition associated with evolutionary changes that occurred since we diverged from chimpanzees about 6 million years ago, in particular the remodeling of the pelvis associated with upright posture.

■ SUMMARY

People think about problems in profoundly different ways. By matching the way you think to the problem you are trying to solve, not only are you choosing the most efficient intellectual tool for the issue at hand, you are also avoiding unnecessary misunderstandings and conflicts. Population thinking and tree thinking are standard in evolutionary biology. In emphasizing the importance of variation in populations, frequencies, probabilities, history, and comparisons, they provide explanations of the origin and state of medical conditions that complement those provided by physics, chemistry, and molecular and cell biology.

What Is a Patient?

Patients are not machines designed by engineers with replaceable parts. They are the result of a long evolutionary process that could only work with the limited variation that existed at the time a change was made. Gradual improvements accumulated in patchwork fashion, each associated with a mutation that usually affected more than one trait, a type of tinkering that resulted in pervasive trade-offs of benefits with costs. If the costs were significant, subsequent changes reduced them, but those changes themselves were part of the same sort of process and resulted in the accumulation of secondary trade-offs. It is astonishing that such an imperfect process could sometimes yield so much precision, complexity, and superb function. It could only do so because the number of selective events was sometimes very, very large.

Virtually every adaptive advance has thus been bought at a cost mediated by a trade-off. The state of a trait is only evolutionarily stable if the associated benefits are greater than the costs. For certain traits, the benefit-to-cost ratio depends on specific circumstances, such as presence or absence of a specific pathogen or food source. In such cases, the trait often evolves to be expressed on demand rather than constitutively. For example, the immune response is very costly, and its benefit, also very high, is only present during an infection. All inducible defenses share this cost-benefit structure: the benefits are contingent, and the defense is produced on demand.

Trade-offs can also be tuned on different timescales. In acclimatization, there is stable, graded physiological adaptation to environmental change, such as ambient temperature or oxygen level. In phenotypic plasticity, the expression of a trait is developmentally regulated by interaction with the environment to produce a single, stable phenotype, such as adult body size. These distinctions, which are basic features of patients, return in Chapter 4 when we discuss defenses. Now, though, we will discuss the nature of patients from a series of perspectives that start with evolutionary events that happened long ago.

Ancient histories with medical consequences

Some important evolutionary events with contemporary medical consequences occurred a *very* long time ago. Asymmetric division of bacterial cells began more than 3 billion years ago; it created the conditions that led to the evolution of aging and the maintenance of the germ line. Multicellularity evolved 1 billion to 2 billion years ago; the stem cells that then evolved to renew and repair tissues in multicellular organisms had characteristics that preadapted them to become cancer cells. About 500 million years ago (mya), a retrovirus inserted into a protoimmunoglobin gene; it made possible the evolution of the vertebrate adaptive immune system. Invasive placentas were found in early mammals but were then suppressed in some lineages. About 15 mya, highly invasive placentas evolved in the shared ancestor of orangutans, gorillas, chimpanzees, and humans. Such placentas are associated with pre-eclampsia—dangerously high maternal blood pressure during pregnancy—and predispose to metastatic cancer. After we diverged from our common ancestor with chimpanzees about 6 mya and evolved bipedal locomotion, the remodeling of the pelvis narrowed the birth canal, leading to special problems with human birth not experienced by other animals.

Let's begin to unpack these ancient events. We will return to each of them in more detail when we discuss aging, cancer, adaptive immunity, parent-offspring conflicts over maternal investment, and reproductive biology.

Asymmetric division: The condition for the evolution of aging

If a cell divides in such a way that one daughter cell gets the new parts and the other the older parts, the daughter lineages in which the older parts accumulate will die out, and those in which the younger parts accumulate will persist. The process has been confirmed in a bacterium, *Escherichia coli* (Figure 2.1). It explains the survival of the germ line since the origin of life, not through impossibly superb molecular maintenance but through continual selection of cells with younger replacement parts, that is, through rejuvenation by sorting out and discarding defects.

If division were perfectly symmetrical, selection would not be able to distinguish between the two daughter cells. Both would be equally intact or equally damaged, and the reproductive payoff from improving the maintenance of either would be equal. As soon as asymmetry causes the reproductive payoff of maintenance in one cell to become smaller, it can be neglected because the other cell has superior reproductive performance. At that point, aging has started to evolve (Weismann 1882; Partridge and Barton 1993). We return to this issue when we discuss the evolution of aging later in this chapter.

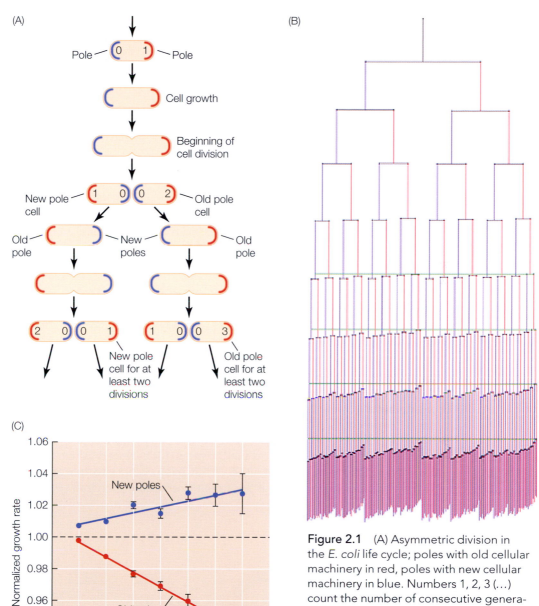

Figure 2.1 (A) Asymmetric division in the *E. coli* life cycle; poles with old cellular machinery in red, poles with new cellular machinery in blue. Numbers 1, 2, 3 (…) count the number of consecutive generations with a given pole type. (B) Cells with new poles divide faster than cells with old poles. (C) Over generations, those differences accumulate. The old cell lines senesce and are replaced by the rejuvenated cell lines. (After Stewart et al. 2005.)

Most cancers originate in stem cells

Stem cells were a great evolutionary innovation that made possible the maintenance of tissues in multicellular organisms. They are involved in important trade-offs, however, because they have characteristics that predispose them to cancer: they retain the potential to differentiate and to move. Stem cells are positioned all over the body, ready to replace cells that wear out and need to be replaced, particularly in bone marrow and the epithelia of lungs, gut, and skin. Those are the tissues in which malignant cancer is most frequent.

The embryonic stem cells that form the invasive placenta are especially well adapted to moving into tissue. Because such cells already contain much of the genetic program needed for metastasis, fewer mutations are needed to produce a cancer. Here traits advantageous to the fetus very early in life carry with them a risk expressed much later in life. Stem cells are thus at the center of a fundamental design trade-off more than a billion years old.

A transposon insertion enabled the vertebrate adaptive immune system

All organisms have defenses against pathogens; we share some of our innate immune system with organisms as ancient as jellyfish. Vertebrates, though, also have an adaptive immune system in which lymphocytes use recombination and mutation to generate the diverse antigen receptors used to identify pathogens. The system is remarkably flexible, capable of sorting among variants to keep pace with pathogens that themselves are rapidly evolving. The machinery that makes this process possible stems from the insertion of a transposable genetic element into an immunoglobulin-like gene about 500 mya in the ancestor of jawed fishes (Cooper and Alder 2006; Murphy et al. 2008). That event built into the gene the mechanism needed to carry out the somatic recombination that generates antibody diversity (Figure 2.2).

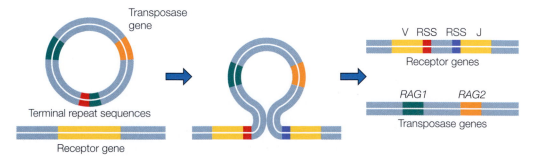

Figure 2.2 The transposase genes needed for somatic recombination in the adaptive immune system were inserted by a mobile genetic element about 500 mya (middle panel) and later separated from their integration site (right panel). V, variable segment; J, joining segment; RSS, recombination signal sequence; *RAG1* and *RAG2*, recombination-activating genes. (After Murphy et al. 2008.)

Invasive placentas occur in lineages with risk of metastasis

When a fetal stem cell invades the endometrium and inserts itself into a maternal spiral artery, it takes partial control of the pipe whose diameter determines the amount of food that the fetus gets. This action is a realization in morphology of a conflict between mother and offspring over the amount of investment that offspring should get, a conflict that is also mediated by hormones produced by placental tissue of fetal origin. It gives the fetus an advantage very early in life that it pays for later in life with increased risk of certain diseases of pregnancy, if female, and of metastatic cancer, often after reproduction has been completed.

In mammals, invasive placentas evolved very early and were later replaced by less invasive placentas in some lineages but not others. The hoofed animals like cows and horses have less invasive placentas, whereas carnivores like cats and dogs have more invasive placentas. Consistent with the idea that metastasis is found in lineages with more invasive placentation, D'Souza and Wagner (2014) found that horses and cows have lower rates of metastatic skin and glandular epithelial cancers and fewer connective tissue sarcomas. The risk of malignant tumors in nonglandular epithelium did not differ among those lineages.

In our hominoid lineage, especially invasive placentas with deep extravillous trophoblast invasion and extensive remodeling of the spiral arteries evolved in the common ancestor of humans, chimpanzees, gorillas, and possibly orangutans (whose placental biology has not yet been well determined) (Figure 2.3). This change occurred about 15 mya and reestablished invasiveness, which evolved in early mammals but was then suppressed in many lineages.

The transition to bipedal locomotion led to problems giving birth

After our lineage split off from the ancestors we shared with chimpanzees, about 6 to 7 mya, it evolved through several forms recognized as distinct

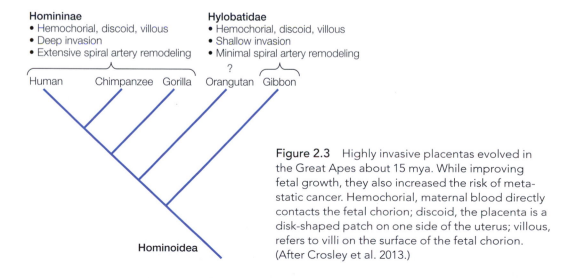

Figure 2.3 Highly invasive placentas evolved in the Great Apes about 15 mya. While improving fetal growth, they also increased the risk of metastatic cancer. Hemochorial, maternal blood directly contacts the fetal chorion; discoid, the placenta is a disk-shaped patch on one side of the uterus; villous, refers to villi on the surface of the fetal chorion. (After Crosley et al. 2013.)

genera: *Sahelanthropus, Ardipithecus, Australopithecus, Paranthropus,* and *Homo*. Bipedal locomotion probably evolved in the earliest of these, *Sahelanthropus*, about 6 to 7 mya, but serious difficulties in childbirth probably came later, in the genus *Homo* about 2.5 mya, as the pelvis continued to be remodeled to support the ability to run down game in cooperative hunts. That hunter, *H. erectus*, also used fire, cooked food, used hand axes, and migrated out of Africa, colonizing Europe and Asia long before our ancestors did and leaving traces in Spain, for example, 800,000 years ago.

■ SUMMARY

Every organism has a history, as does every trait. Traits have different ages of origin, and organisms consist of parts that have existed for different amounts of time. The older the trait, the more it has been embedded in development, and the harder it probably is to change it. Things that evolve slowly constrain the evolution of things that could otherwise evolve more rapidly. Some traits of medical significance are billions of years old, and patients are mosaics of traits with different ages, traits assembled by evolution in a series that started with metabolism at the origin of life. Features such as the following have been added in a long sequence that will continue into the future:

• Metabolism	3900–1600 mya
• Eukaryotic cells, chromosomes, and sex	2500–1600 mya
• Development and physiology (multicellularity)	ca. 1000 mya
• The vertebrate immune system	ca. 500 mya
• Endothermy and the scrotum	ca. 230 mya
• Lactation	ca. 230 mya
• The placenta	ca. 190 mya
• Highly invasive placentas	ca. 15 mya
• Bipedalism	ca. 7 mya
• Short interbirth interval and sharing food and child care	<5 mya
• Hairless, dark skin color, and abundant sweat glands	2–3 mya

Recent history generating diversity

Traits that humans do not share with chimpanzees and bonobos include delayed maturation, menopause, and longevity as well as those listed in Table 2.1. Some of the advantages in Table 2.1 were accompanied by costs. Prolonged brain growth required a higher percentage of body fat in the newborn to support the brain, made success in infant development more sensitive to poor nutrition, and made reliable child care essential. The evolution of sweat glands was accompanied by loss of hair to make cooling more effective during endurance running, which increased the risk of melanoma. The earlier onset and longer duration of labor compensated partially for the mismatch between infant head circumference and birth canal but increased the risk of damage to the infant brain through oxygen deprivation. The improvement in emotional communication made possible

---

TABLE 2.1   Recent changes in human evolution

Example of phenotypic feature	Human lineage–specific trait	Possible evolutionary advantages
Brain growth trajectory	Prolonged postnatal brain growth and delayed myelinization period; enhanced cognition	Allowed creation of novel solutions to survival threats; increased the critical period for learning new skills; facilitated emergence of uniquely human cognitive skills
Brain size	Increased brain/body size ratio; enhanced cognition	Allowed creation of novel solutions to survival threats; improved social cognition
Descended larynx	Portion of tongue resides in throat at level of pharynx; larynx descended into throat	Helped develop spoken languages
Eccrine sweat gland density	Higher density of eccrine glands; enhanced sweating capacity	Enhanced cooling ability; allowed protection of heat-sensitive tissues (e.g., the brain) against thermal stress; facilitated endurance running
Endurance running	Improved energy use during periods of high energy demand; increased capacity to transfer energy (in the form of glycerol) from fat stores to muscle; anatomical changes relating to running ability	Allowed persistence hunting to emerge as a viable strategy for accessing the benefits of increased meat consumption; increased range of food sources; improved diet may have facilitated brain evolution
Labor	Earlier onset and longer duration of labor	Partially protected the child and mother from damage due to increased head circumference
Lacrimation	Emotional lacrimation (crying)	Enhanced emotional communication within social groups; increased affective communication
T-cell function	Relative T-cell hyperreactivity	Enhanced immune function
Thumb	Increased length, more distally placed; larger associated muscles	Allowed creation of more detailed tools; allowed manipulation of objects on a finer scale

Source: After O'Bleness et al. 2012, Box 1, pp. 2–3.

by the origin of crying was probably accompanied by greater susceptibility to emotional disorders and predator attacks. Our recent evolution was not free of trade-offs.

Migration out of Africa

We were preceded out of Africa by *H. erectus*, which colonized Europe and Asia more than a million years ago. Modern humans, *H. sapiens*, evolved in Africa about 150,000–250,000 years ago and migrated out of Africa about 100,000 years ago (Figure 2.4). We know that we are all Africans because the fossil and genetic evidence strongly supports that conclusion. As we migrated across the planet, groups encountered different diseases and

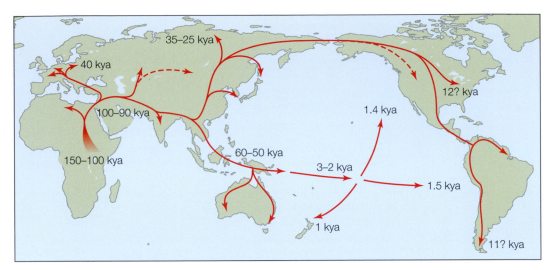

Figure 2.4 *Homo sapiens* colonized the planet relatively recently. Kya, thousand years ago. (After Barbujani and Excoffier 1999.)

diets and responded to them with genetic changes that can still be seen as signatures of selection in the genome. The groups that moved into America and Polynesia were small enough to leave some infectious diseases behind. One result has been great genetic variation among humans in ability to resist disease and metabolize drugs.

Genetic evidence on recent human evolution

To put what follows into context, note that all humans are, from one point of view, nearly identical genetically. Our genome is so large that both of the following statements are true: I am genetically 99.9% identical to you (we share 99.9% of 3.3 billion nucleotides), but my genome differs from yours at up to 3.3 million positions (0.1% of 3.3 billion nucleotides). The first suggests how conservative inheritance is, and the second points to its flexibility. Note that nucleotide differences are not all equivalent. Some are neutral, some alter the structure of a single protein with strictly local effects, and some change the control over large genetic networks. Genetic differences among human individuals are of these types and more.

The sequencing of entire human genomes has led to the discovery of millions of single nucleotide polymorphisms (SNPs), places at which two strands of human DNA differ at a single nucleotide site. Because many of them are very rare and therefore not helpful for studying history and relationships, it is required by convention that the set of SNPs used have minor allele frequencies of 0.01 or greater (the second most common allele must be present at least 1% of the time). A comprehensive look at human genetic diversity was taken by Li and colleagues (2008), who analyzed 650,000 SNPs from 938 unrelated individuals drawn from 51 populations. They were able

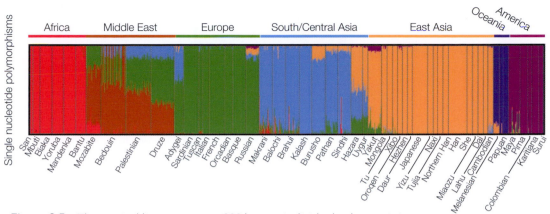

Figure 2.5 The vertical lines represent 938 human individuals, the x-axis is labeled with the populations from which they were sampled, the y-axis represents 650,000 SNPs, and the colors represent the groups inferred from a grouping algorithm set in this case to recognize seven groups. (From Li et al. 2008.)

to detect individual ancestry and population substructure with very high resolution and express it in a single picture of human genetic variation (Figure 2.5). That picture reveals that the human populations of Africa, Europe, East Asia, and America are more genetically homogeneous than are the populations of the Middle East and South/Central Asia, which contain more individuals with a recent mixture of genetic ancestries.

This detailed genetic information was used to reconstruct the relationships of the 51 populations. That reconstruction (Figure 2.6) demonstrates our African origin quite clearly. The long branch leading from the African to the Middle Eastern populations is a signature of a genetic bottleneck associated with the relatively small group of people who moved out of Africa, leaving a great deal of genetic diversity behind. Similar bottlenecks can be seen in the long branches leading to the Oceanic populations and the Native Americans. The timing of movements is important for several reasons. Our ancestors moved into Eurasia in time to encounter and hybridize with Neanderthals and Denisovans, who disappeared more than 30,000 years ago, and their efficiency as hunters is a probable explanation for the extinction of the Pleistocene megafauna of both Eurasia and the Americas. The last mammoths died out 4500–6500 years ago.

Some of the genetic diversity of modern humans is due to adaptations to different environments as humans spread across the globe, and some is due to founder effects and genetic drift. The migrants leaving Africa carried with them only a small sample of the genetic variation that then existed in Africa, and the migrants to North America and Oceania repeated that sampling process. Modern human genomes therefore have different mixes of selected and random variation. Both types of genetic diversity can affect disease susceptibility, but the patterns are likely to be different. Selected variation provided an advantage in past environments and may

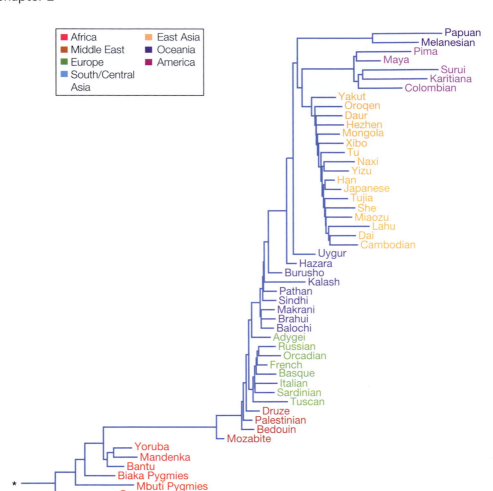

Figure 2.6 A detailed look at global human genetic variation establishes our African origin and a sequence of population relationships, confirming the map in Figure 2.4, which was inferred from archaeological and linguistic evidence. The populations and color-coded locations in this phylogenetic tree match those shown in Figure 2.5. (After Li et al. 2008.)

either still be adaptive, may be neutral, or may have become detrimental. For example, increased resistance to some infections may predispose to autoimmune diseases. Genetic diversity caused by drift, on the other hand, was not selected and may remain neutral or become adaptive or maladaptive in a less predictable fashion.

Support for the concept of race is weak

A sample as large as Li and colleagues' (2008) 650,000 measurements on 938 individuals gives great statistical power to discriminate among small differences. Many of those measurements are correlated with one another, which in this case means that SNPs tend to vary among individuals in similar patterns. To remove those correlations and reduce the large number of measurements to a few independent explanatory variables, one performs a principle components analysis. The result of such an analysis is a set of principle components, the first of which (PC1) captures the most variation in the data, the second (PC2) the second most, and so forth. Principle components are calculated so that the variation measured by one principle component is not correlated with the variation measured by the others. When a principle component analysis was performed on the European populations in this data set, the first two principle components discriminated clearly among the Sardinians, Basques, Russians, and Adygei (from the foothills of the Caucasus), which each formed a group around a central cluster that contained the French, Italians, Tuscans, and Orcadians (from the Orkney Islands), which were less clearly discriminated (Figure 2.7).

Note that PC1 explains 2.4% of the variation in this data set; PC2 explains 1.6%. Thus, only 4% of the genetic variation detected in Europe is represented in an analysis that discriminates the populations from which the individuals were sampled.

That raises a key question: How much of human genetic variation is among individuals, how much among populations, and how much among ethnicities? Four studies have answered that question, and their results are in strong agreement. If we sample individuals from many populations and ethnicities, we find that 84%–89% of their genetic variation arises from differences among individuals, 2%–8% from differences among populations, and 6%–11% from differences among ethnicities or geographic regions;

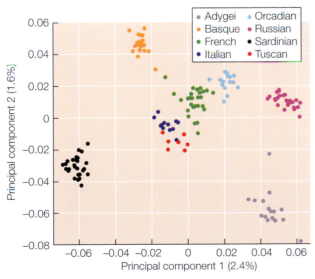

Figure 2.7 Principle components analysis of genetic variation among European populations. One point indicates one individual. (After Lewontin 1995; Li et al. 2008.)

TABLE 2.2 Human genetic variation

Polymorphism	Number of loci	Number of groups	Within populations/ samples	Among populations/ samples	Among groups
Protein	17	7	84.5	8.3	6.3
Protein	25	3	86	2.8	11.2
DNA	109	4–5	84.4	4.7	10.8
SNPs	650,000	7	88.9	2.1	9

Sources: Protein (first): Lewontin 1995; protein (second): Ryman, Chakraborty, and Nei 1983; DNA: Barbujani et al. 1997; SNPs: Li et al. 2008.

see Table 2.2. Thus, most human genetic variation is accounted for by differences among individuals, and biological support for the concept of race is weak.

The concept of race only explains at most about 10% of the genetic differences between two individuals chosen at random. Skin color and facial morphology are superficial and misleading indicators of biological differences. Our primate ancestors had light skin under dark hair. Their exposed skin was light at birth and darkened with age. About 2 to 3 mya, when we lost our body hair to improve the efficiency of sweating in regulating our body temperature as long-distance runners who ran down prey, we evolved black skin to protect against skin cancer and folate depletion. Between 100,000 and 10,000 years ago, some human populations evolved lighter skin colors to improve vitamin D synthesis in the recently colonized cool climates where people wore more clothing (Jablonski and Chaplin 2010). Coded by a small number of genes, skin color is caught in a trade-off between the benefits of vitamin D, the risks of skin cancer, and the costs of folate depletion, which occurs more rapidly in people with light skin colors.

■ SUMMARY

Many traits of medical significance have experienced evolutionary change since we shared ancestors with chimpanzees and bonobos. We emerged from Africa about 100,000 years ago. As we moved into environments with different diets and diseases, we evolved local genetic differences in disease resistance and drug metabolism. Biological support for the concept of race is weak; race is primarily a social construct. Because of the huge size of the genome, however, there are genetic differences associated with ethnicity that contain useful medical information about drug metabolism and disease resistance. We discuss some of them below.

Variation in disease resistance

The idea that humans vary genetically in their ability to resist disease was suggested by at least three observations: children are at greater risk of succumbing to an infectious disease if one of their parents has had it; there are several diseases, like tuberculosis and bacterial meningitis, where many are infected but few get sick; and people living in areas where a disease is endemic are often more resistant than those coming from areas where that disease does not exist. The study of genetic disease resistance began with candidate gene approaches, which used a combination of biological insight, biochemistry, and cell biology to identify the hemoglobin variants that cause sickle cell anemia. More recently, case-control genome-wide association studies (GWAS) have suggested many more candidate genes, and whole-exome sequencing is now spreading a wider net more finely to catch rarer variants. Of the hundreds of infectious diseases that afflict humans, only the few discussed next have been shown to interact significantly with human genetic variation.

Genome-wide association studies (GWAS) using single nucleotide polymorphisms (SNPs)

GWAS rely on SNPs used as markers for regions of chromosomes that contain several to many genes coding for proteins or control elements. The basic idea is that a single nucleotide position is identified as variable and marks a region of a chromosome containing genes that may also vary among individuals because it is linked to those genes by proximity. Sometimes, but not always, the SNP is actually located in a gene; more frequently, it is in a region of DNA that is close to genes but not expressed as a protein. Case-control studies contrast the versions of genes found in people with a disease (the cases) with those found in people who do not have the disease (the controls). Those versions are identified by the SNP variants marking the chromosomal regions containing the genes.

Importantly, only SNPs whose minor allele frequencies are greater than 1% or 5%—depending on the platform—are included in the analysis. That means that some chromosomal regions with rare variations will be missed. That might not seem problematic until one realizes that the genome is so large that the sum of many effects, each of them rare individually, could be significant in total.

GWAS scan the entire genome for signals of significant association of chromosomal regions with disease risk. If the persons studied have been sampled for 650,000 SNPs, 650,000 statistical tests are performed, one for each SNP. Whenever many tests are performed, there is some chance that a significant result will occur at random. For that reason, GWAS now require conservative probability thresholds to accept a result as potentially significant: $p < 5 \times 10^{-8}$ is often used—that is, 0.05, the conventional significance threshold, divided by 1 million—to correct for performing 1 million tests. Usually, several thousand cases and several thousand controls are needed to yield credible results at that significance level. It has become standard to require that candidate genes so identified have independent confirmation from a similarly powerful study because many early studies that did not use stringent probability thresholds and were based on smaller samples produced results that could not be replicated.

TABLE 2.3 Six genetic variants that increase risk of infectious diseases

Genetic variant/condition	Gene	Infectious disease	Year reported
Sickle hemoglobin	*HBB*	*Plasmodium falciparum* malaria	1954
Duffy blood group	*DARC*	*P. vivax* malaria	1976
Prion protein gene variant	*PRNP*	Creutzfeldt-Jakob disease	1991
Melanesian ovalocytosis	*SLC4A1*	*P. falciparum* malaria	1995
CC chemokine receptor-5 Δ32	*CCR5*	HIV-1 infection	1996
Blood group nonsecretion	*FUT2*	Norwalk virus diarrhea	2003

Source: After Hill 2012, Table 1, p. 842.

Sequencing approaches

GWAS have identified many candidate genes, some of which we will see below, but those identified only account for a small percentage of the heritable variation in disease resistance. This situation is called the paradox of the missing heritability. Where are the missing effects located? One possibility is that they are caused by many rare alleles, each of which has a large effect. The rare alleles are not captured by a SNP analysis that only includes variants with frequencies greater than 1% or 5%, but they could be detected in a whole-genome sequence. Sequencing the entire genome has become much cheaper than it once was, but at about $1000 per person it remains very expensive to do a study that needs a sample of several thousand people. Therefore, until sequencing becomes much cheaper, researchers are starting to sample whole exomes, which constitute about 1% of the genome. The exome, which consists of all exons—that is, all the DNA that is translated into proteins—does not include all genetic control elements, but it does include all protein variants.

Genes with large effects

Hill (2012) discussed six cases in which genes were identified using biological information prior to GWAS (Table 2.3). They all have large effects, with an odds ratio (OR) greater than 3. The OR measures the risk of a condition compared to a control; thus, OR > 3 means that a person with one genetic variant (usually one homozygote) is at least three times more likely to get the disease than a person with the other genetic variant (the other homozygote). Hill found that all six genetic variants have substantial effects on the risk of an important infectious disease and are present at substantial frequency in some human population. The early analyses of resistance to malaria mediated by sickle hemoglobin (*HBB*) and the Duffy blood group (*DARC*) and to Norwalk virus diarrhea mediated by nonsecretion of blood-group antigens (*FUT2*) all included studies of deliberately infected volunteers as opposed to population samples of cases and controls, which were used to identify the remaining three genes.

Genes with moderate effects

If we include with the 6 genes studied in Table 2.3 those with moderate effects, 0.5 < OR < 2—that is, genes that halve or double the risk of disease—we get another 15 genes, some identified with candidate gene approaches and some with GWAS. These effects are stronger than those usually found with GWAS. Two things are especially striking about the genes listed in Table 2.4. First, as many genetic variants have been found that protect against leprosy (4 variants) as against malaria. Evidently, both diseases have played an especially important role in our evolutionary history. Second, one might question whether a whole-exome analysis would discover many disease genes because it only includes 1% of the genome and ignores many genetic control elements. In this sample, however, 16 of 17 variants would have been identified in whole-exome analyses, which therefore might not miss many important genes. Thus, in this study, GWAS confirm the effects of several genetic variants identified by earlier approaches and add to the list about 15 variants with moderate effects (0.5 < OR < 2.0) (Hill 2012).

TABLE 2.4 Loci strongly associated with infectious disease susceptibility

Gene	Disease	GWAS?	Exomic?	Minor allele
Hemoglobin S	Malaria		Yes	Protective
SLC4A1 (ovalocytosis)	Malaria		Yes	Protective
CCR5	HIV/AIDS		Yes	Protective
PRPN	vCJD		Yes	Protective
FUT2	Norovirus		Yes	Protective
Duffy blood group	Vivax malaria		No	Protective
HLA-DR/DQ	Leprosy	Confirmatory	Yes	Both
HLA-B	HIV/AIDS	Confirmatory	Yes	Both
HLA-DQ/DP	HBV	Yes	Yes	Both
HLA-C	HIV/AIDS	Confirmatory	Yes	Both
Blood group O	Malaria		Yes	Protective
G6PD	Malaria		Yes	Protective
CFH	Meningococcus	Confirmatory	Yes	Protective
MAL	Bacteremia		Yes	Protective
TLRI	Leprosy	Confirmatory	Yes	Protective
IL-28B	HCV	Yes	Yes	Susceptible
MBL2	Pneumococcus		Yes	Susceptible
C13orf31	Leprosy	Yes		Susceptible
CCDC122	Leprosy	Yes		Protective

Source: After Hill 2012, Table 2, p. 842

TABLE 2.5 Genetic variants with significant effects, some relatively small, identified by GWAS

Disease	Phenotype	Population	Sample size
HIV-1 and AIDS	Viral load at set point	European	2554
	Viral load at set point	African American	515
	HIV-1 control	European	1712
		African American	1233
	Disease progression	European	1071
	Progression to AIDS 1987	European American	755
	Long-term nonprogression	European	1627
	Long-term nonprogression	European	1911
Hepatitis C	Spontaneous clearance	European	1362
Hepatitis B	Chronic infection	Japanese, Taiwanese	6387
Dengue	Dengue shock syndrome	Vietnamese	8697
Severe malaria	Susceptibility	African (Gambian)	5900
Tuberculosis	Susceptibility	African (Ghana, Gambia, Malawi)	11,425
Leprosy	Susceptibility	Chinese	11,140
Meningococcal disease	Protection	European	7522
Variant Creutzfeldt-Jakob disease	Susceptibility	European, Papua New Guinea	5183

Source: After Chapman and Hill 2012, Table 1, p. 177.

Genes with small effects

GWAS also pick up a much larger set of candidate genes with smaller odds ratios (Table 2.5). Note that these variants, and their effects, are specific to populations found in geographical regions where the relevant diseases are

Most significant marker(s)	SNP location	p-value	Odds ratio
rs9264942	HLA-C	5.9×10^{-32}	—
rs2395029	HLA-B, HCP5	4.5×10^{-35}	—
rs2523608	HLA-B	5.6×10^{-10}	—
rs9264942	HLA-C	2.8×10^{-35}	2.9
rs4418214	MICA	1.4×10^{-34}	4.4
rs2395029	HLA-B, HCP5	9.7×10^{-26}	5.3
rs3131018	PSORS1C3	4.2×10^{-16}	2.1
rs2523608	HLA-B	8.9×10^{-20}	2.6
rs2255221	Intergenic	3.5×10^{-14}	2.7
rs2523590	HLA-B	1.7×10^{-13}	2.4
rs9262632	Intergenic	1.0×10^{-8}	3.1
rs9261174	ZNRD1, RNF39	1.8×10^{-8}	—
rs11884476	PARD3B	3.4×10^{-9}	—
rs2395029	HLA-B, HCP5	6.8×10^{-10}	3.47
rs2234358	CXCR6	9.7×10^{-10}	1.85
rs8099917	IL28B	6.1×10^{-9}	2.31
rs3077	HLA-DPA1	2.3×10^{-38}	0.56
rs9277535	HLA-DPB1	6.3×10^{-39}	0.57
rs3132468	MICB	4.4×10^{-11}	1.34
rs3765524	PLCE1	3.1×10^{-10}	0.80
rs11036238	HBB	3.7×10^{-11}	0.63
rs4334126	18q11.2 (GATA6, CTAGE1, RBBP8, CABLES1)	6.8×10^{-9}	1.19
rs3764147	LACC1	3.7×10^{-54}	1.68
rs9302752	NOD2	3.8×10^{-40}	1.59
rs3088362	CCDC122	1.4×10^{-31}	1.52
rs602875	HLA-DR-DQ	5.4×10^{-27}	0.67
rs6478108	TNFSF15	3.4×10^{-21}	1.37
rs42490	RIPK2	1.4×10^{-16}	0.76
rs1065489	CFH	2.2×10^{-11}	0.64
rs426736	CFHR3	4.6×10^{-13}	0.63
rs1799990	PRNP	2.0×10^{-27}	—

endemic. The sample sizes are large, and the p-values are very small; many other potential candidate genes were not included on this list because they did not make it through the statistical screen. The diseases for which such genetic variants have been identified include AIDS, hepatitis C and B,

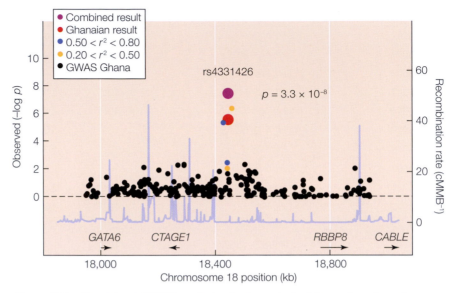

Figure 2.8 Combined GWAS analysis of tuberculosis risk as a function of genetic variation in Gambia and Ghana. (After Hill 2012.)

dengue, malaria, tuberculosis, leprosy, meningococcal disease, and variant Creutzfeldt-Jakob disease (prion-mediated).

Let's take a closer look at the results for one disease, tuberculosis, in one place, West Africa. In Figure 2.8, they are displayed in a plot with the negative logarithm of the p-value on the left y-axis (where a value of 8 means a probability of observing that association at random of 10^{-8}) and the chromosomal position on the x-axis. The significant SNP (rs4331426) is not in a gene, but it is close to, and serves as a marker for, four genes (*GATA6*, *CTAGE1*, *RBBP8*, and *CABLE*) on chromosome 18 that are within 500 kilobases (kb). The recombination rates along that portion of the chromosome are plotted (underneath, in blue), with the scale on the right y-axis. The gene most tightly linked to the SNP is *RBBP8* because although *CTAGE1* is physically closer on the chromosome, more recombination goes on between it and the SNP. Note that in this case, as in many, the SNP found to be significant in the GWAS was not in a gene but was associated with several nearby genes, any or all of which could be mediating the disease resistance. GWAS identify candidates by correlational analyses that have to be confirmed by functional studies capable of identifying causes.

The effect of variation in the pathogen

It is striking that so many variants have been identified that influence the risk of getting leprosy (see Table 2.5), and it is perhaps not a coincidence that the leprosy pathogen, *Mycobacterium leprae*, is less genetically variable than most. It is thus possible that the yield of candidate genes would be greater if the GWAS were stratified by genetic strain of pathogen.

Maintaining sample sizes large enough to ensure statistical power with such an approach would be a challenge.

■ SUMMARY

Humans vary genetically in their ability to resist diseases, but for only a few genes and a few diseases has a causal connection been convincingly identified. Those diseases include malaria, AIDS, leprosy, and bacterial meningitis. The GWAS approach that searches for common variants with common effects has had some success but has not lived up to initial hopes because the sum of their effects does not add up—by far—to the heritability of resistance observed in phenotypes. The search is now turning to rare, highly penetrant variants in single genes and to rare, diverse variants in multiple genes that are not fully penetrant. Both types can be found with large samples of whole-exome sequences and, even better, whole-genome sequences (Hill 2012).

Variation in drug metabolism

Because of their different evolutionary histories, humans vary genetically in their ability to metabolize drugs. Thus, physicians need to be concerned about getting the dose correct, for example, when prescribing methadone and antidepressants because doses that work well for one patient may be ineffective for others. Cells that mutate to form cancers bring with them the genetic variation in the ability to metabolize drugs that they inherit from the germ line; therefore, patients also vary for genetic reasons in their responses to chemotherapy. In addition, some patients, a small minority, are at risk of having a serious adverse drug reaction that could lead to heart failure, liver transplant, or paralysis. For all these reasons, physicians need to be sensitive to the genetic variation in their patient populations.

It was adverse drug reactions that first alerted the medical community to these issues. They were of two types. Type A reactions were based on known pharmacology, not on genetic information, and were predictable, relatively frequent, and seldom fatal. Thalidomide, which caused serious birth defects, was the leading example. Type B reactions were, at the time, unpredictable, idiosyncratic, usually infrequent, and more often serious or fatal (Venning 1983). They included heart attacks, kidney failure, colitis, and anemia. Over the 33 years from 1969 to 2002, 2.3 million cases of adverse reactions to about six thousand drugs were reported to the U.S. Food and Drug Administration. More than 75 drugs were removed from the market, 11 were given special restrictions, and less than 1% were restricted or withdrawn (Wysowski and Swartz 2005); most of those involved type A reactions.

Many of the idiosyncratic type B reactions were caused by individual and interethnic variation in cytochrome P450s (CYPs) and N-acetyl transferases (NATs), the products of two gene families involved in metabolizing the majority of drugs. Those gene families are ancient. They evolved to process toxins in food—many of them chemicals that plants evolved for defense against herbivores—and in smoke and other aerosols.

Figure 2.9 Genetic variation is only one reason patients vary in drug response. (After Meyer, Zanger, and Schwab 2013.)

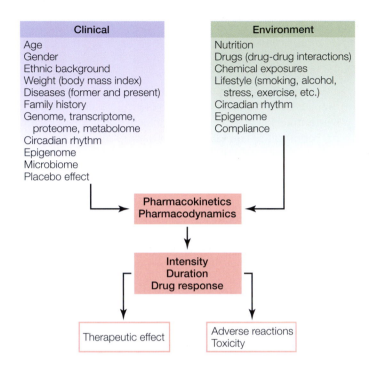

Genetic variation, however, is not the only reason patients vary in drug response; many environmental and clinical factors also influence the efficacy and safety of drugs (Figure 2.9). In many cases, modern humans are using this ancient metabolic machinery to respond to synthetic compounds of which they have had no previous evolutionary experience, and such responses cannot be expected to be optimal.

Variation in cytochrome P450s (CYPs)

CYPs are an ancient set of families of enzymes that evolved by repeated gene duplications. They are proteins bound to a heme group, which gives them their color in solution, with an absorption peak at 450 nm that accounts for their name. Humans have 57 CYP genes whose products bind to the endoplasmic reticulum and the inner mitochondrial membrane, where they mediate both toxin and steroid metabolism as well as other reactions. The products of 23 genes in three families account for about 75% of human drug metabolism. Living organisms use heme groups to manipulate oxygen precisely, and CYPs catalyze oxidative reactions.

We focus on one of them, *CYP2D6*, whose genetic variation, with many identified alleles, is clinically important. That variation was initially inferred in 1977 in London from a volunteer's hypotensive response to debrisoquine, an antihypertensive, and in Bonn from side effects in a patient taking sparteine, an antiarrhythmic (Meyer and Zanger 1997). There is considerable ethnic variation in the frequency of *CYP2D6* alleles: *1, *2, *9, *10, and *17

TABLE 2.6 Ethnic variation of the *CYP2D6* alleles

	n	*1	*2	*3	*4	*5	*6	*9	*10	*17
Asians										
Chinese	226	26.9	13.4			5.7			50.7	
Japanese	412	43.0	12.3		0.2	4.5			38.1	
Korean	304	49.0							51.0	
Blacks										
American	492	83.0		0.6	7.3	6.9			5.2	26.0
Tanzanian	216	27.8	40.0	0.0	0.9	6.3	0.0		3.8	17.0
Zimbabwean	160	85.6	9.9	0.0	2.5	3.8		0.0	5.6	34.0
Whites										
American	928	76.0		1.2	18.1	2.9			4.0	0.0
British	1332	33.4	32.9	1.8	18.9	7.3	1.4	2.6	1.4	0.1
German	1154	36.4	32.4	2.0	20.7	2.0	0.9	1.8	1.5	

Source: After Xie et al. 2001, Table 3, p. 822.

Note: n = number of individuals sampled; frequencies are expressed as percent. Allelic variants are written *CYP2D6*1*.

are functional, and the rest are not (Table 2.6). *CYP2D6* is representative in several senses: it metabolizes many drugs, it varies genetically, its alleles are found at different frequencies in different ethnicities, and its ethnic variation is sufficient to alert a physician to a potential issue but is not a reliable prediction for drug prescription because there is significant intraethnic variation.

The genetic variation in *CYP2D6* is not only allelic; it also varies in number of copies from 2 to 13 per individual. It metabolizes 25% of the drugs on the market, and its polymorphisms influence the processing of antidepressants, antiarrhythmics, anticancer drugs, and analgesics. Its duplication reduces the efficacy of drugs that treat arrhythmia, Alzheimer's disease, and heroin addiction but improves Tamoxifen treatment of breast cancer. The drug industry now tries to avoid developing drugs that are metabolized by *CYP2D6* because it is so difficult to predict how individuals will react to it.

Variation in the products of other CYP genes has dramatic effects on the efficacy of drug rehabilitation. Genetic polymorphisms in *CYP3A4* and *CYP2B6* affect methadone metabolism so strongly that to obtain plasma concentrations of 250 ng/mL, doses as low as 55 mg/day or as high as 921 mg/day can be required in a 70 kg patient, a more than 16-fold variation. The appropriate dose can be determined by acquiring the relevant genetic information.

Variation in N-acetyl transferases

Two *N*-acetyl transferases, NAT1 and NAT2, have clinically important individual variation. They both activate and deactivate drugs and carcinogens in the liver cytosol. Their alleles combine to produce rapid, intermediate, and slow acetylator phenotypes. Initially discovered in 1953 in patients being treated for tuberculosis with isoniazid, they metabolize sulfonamides, caffeine, and many other drugs. Like CYPs, NATs display considerable variation among ethnicities: whereas Europeans and North Americans are 22%–26% fast acetylators, East Asians are 67%–74% fast acetylators (Meyer and Zanger 1997).

NATs mediate cancer risk in two ways: they process environmental toxins related to cancer risk, and they process drugs used in chemotherapy. They are thus at the center of key genotype × environment interactions affecting cancer incidence and treatment. For example, NAT2 slow acetylators are 1.46 times more likely to get bladder cancer, and those with the combination of NAT2 slow acetylator, NAT1 fast acetylator, and a history of cigarette smoking are 2.73 times more likely to get it (Sanderson et al. 2007). NAT1 fast acetylators have a significantly higher risk of colorectal cancer than do slow acetylators, and smoking increases risk in carriers of both NAT1 and NAT2 fast acetylators (Sørensen et al. 2008).

NAT1 polymorphisms are associated with the risk of pancreatic and lung cancer as well as myeloma. NAT2 polymorphisms are associated with the risk of non-Hodgkin lymphoma, liver cancer, colorectal cancer, and bladder cancer (Agundez 2008).

Individual reactions to chemotherapy

All the cells in a cancer share some genetic capacity to process drugs with that patient's healthy cells, which began life with potential responses to chemotherapy inherited from the germ line. In addition, cancer cells accumulate many somatic mutations, some of which may affect the ability to metabolize drugs and thus resist chemotherapy. When chemotherapy applies strong selection to competing clones of cancer cells and rapidly selects resistant clones, some of that resistance is probably attributable to mutations in genes coding for CYPs and NATs.

Using GWAS to discover genetic variation for drug response

Genes involved in adverse reactions to widely used drugs have been identified with GWAS (Daly 2010). The first study involved the cholesterol-lowering drug simvastatin, which in some patients causes the breakdown of muscle tissue. The contrast of 85 cases with 90 controls (patients who took the drug but had no side effects) for 310,000 SNPs located a gene, SLCO1B1, that expresses a solute carrier organic anion transporter. The variant allele of that gene increased the risk of myopathy 4.5-fold per copy carried. The second study involved liver damage in some patients treated with an antimicrobial, flucloxacillin, for staphylococcus infections. The contrast of 51 cases with 282 controls for 1 million SNPs located a gene in the major histocompatibility complex whose variation was associated with 80-fold increased risk of liver damage.

TABLE 2.7 Genome-wide association studies on variation in drug response

Clinical problem	Number of published studies	Drugs involved	Gene variant(s) tagged by SNP
DILI	4	Ximelagatran, Flucloxacillin, Lumiracoxib, Amoxicillin-clavulanate	HLA class I and II variants
Hypersensitivity reaction (SJS, TEN)	3	Carbamazepine	*HLA-B*1502, HLA-B*3101*
Myotoxicity	1	Simvastatin	*SLCO1B1*
Lack of efficacy	1	Clopidogrel	*CYP2C19*
Efficacy of treatment of HCV infection (viral RNA in serum)	3	Peginterferon-α	*IL28B*
Variable individual dose requirement of coumarin anticoagulants	3	Warfarin, Acenocoumarol, Phenprocoumon	*VKORC1, CYP2C9, CYP4F2*
Individual variability in glycemic response in type 2 diabetes	1	Metformin	*ATM* (?)
Individual variability in clinical outcomes in breast cancer	1	Tamoxifen	*C10orf11*
Individual variability in clearance and toxicity	1	Methotrexate	*SLCO1B1*

Source: After Meyer, Zanger, and Schwab 2013, Table 3, p. 482.
Note: Only highly significant results with replication are reported.

Since then, the list of genes involved in adverse drug responses identified through GWAS has rapidly grown (Table 2.7). It is not clear why GWAS have been more effective at locating genes involved in adverse drug responses than they have been at locating genes involved in disease resistance (Meyer, Zanger, and Schwab 2013).

■ SUMMARY

Patients vary genetically, both in alleles and in copy number, in their ability to process drugs and environmental toxins. If a patient cannot process a drug molecule fast enough, its concentration can rise to toxic levels. Cancer cells also vary in their sensitivity to drugs, which accounts for some of the evolution of resistance to chemotherapy. Knowing the constitution of individual patients and individual cancer clones for genes involved in drug metabolism can improve treatment outcomes and reduce the risk of adverse drug reactions.

Variation in life history traits

A patient is not just the snapshot in time that presents in the physician's office or hospital bed. A patient has an entire life history that has accumulated from the fertilization of the zygote through birth and growth and is a complex and layered historical accumulation of genotype × environment interactions. While they develop in the course of each individual's life, life histories also evolved and continue to evolve. In many ways, the human life history is derived, special, and unique. Because it is starting to change under selection caused by medical and public health measures, it now has medical causes as well as medical consequences.

While the focus in much of evolution is on genetic change, life history evolution focuses on phenotypes. Patients are phenotypes, phenotypes are composed of traits, and doctors usually treat traits, not genes (with the exception of gene therapy). In analyzing trait evolution, life history theory has successfully explained, to a first approximation, the major life history traits: why organisms are large or small, why they mature early or late, why they have few or many offspring, why they have a short or a long life, and why they must grow old and die.

To do this analysis, life history theory framed the evolution of life histories as an interaction between extrinsic and intrinsic factors. The extrinsic factors influence age-specific rates of mortality and reproduction. That is where the environment enters in the form of nutrition, disease, violence, medicine, and public health. The intrinsic factors are captured in trade-offs among traits, relationships in which changes to one trait that increase reproductive success are linked by whatever mechanism—genetic, developmental, morphological, or physiological—to changes in other traits that decrease reproductive success. This approach thus focuses on the analysis of costs and benefits: how much can one trait change to improve fitness before costly changes in other traits cancel those benefits? For example, what are the costs and benefits of maturing earlier or later, or of investing more in each of a few offspring, or less in each of many of them? Is there an optimum with the best outcome?

This method of analyzing trait evolution, developed for life history traits, suggests how all traits evolve. Now let's see how it works for one key trait, the maturation event.

Age and size at maturity

The maturation event is a milestone in the life history of any individual because it marks major shifts in allocations among growth, maintenance, and reproduction. Those shifts initiate aging as well as adulthood. Natural selection shapes the evolution of the maturation event to maximize the benefits of maturing at a given age and size. One analysis makes these assumptions about the key trade-offs: organisms that mature later produce offspring that survive better, organisms that mature later grow longer and can produce more offspring because of their larger size, earlier maturity shortens generation time, and earlier maturity shortens the period of

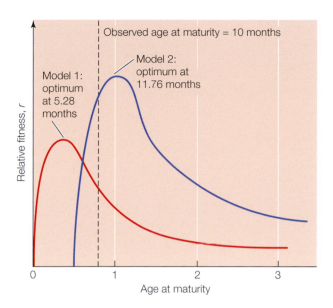

Figure 2.10 Fitness profiles depicting the evolutionary consequences of maturing at different ages in the western fence lizard. Model 1 assumes only that juvenile mortality declines with age at maturity; Model 2 assumes only that fecundity increases linearly with size and that size increases with age. (After Stearns and Crandall 1981.)

risk before reproduction begins. Those assumptions define relationships among ages, x; age-specific survival rates, $l(x)$; age-specific fecundity rates, $m(x)$; and ages at maturity (α) and last reproduction (ω). The Euler-Lotka equation relates them all to fitness (r):

$$1= \int_{\alpha}^{\omega} {}^{-rs}l(x)m(x)dx$$

The trade-offs are built into the equation by substituting functions for the survival and fecundity terms. One then asks, what value of age at maturity (α) maximizes fitness (r) given those trade-offs? The situation faced by organisms in making the decision to mature can be unpacked by plotting fitness against age at maturity as done in Figure 2.10 for the western fence lizard.

The analysis in Figure 2.10 used estimates of the relationships between fecundity and age and between juvenile mortality and age from empirical observations. It makes these general points: maturing too early is a bad idea because waiting will improve both the quality and the quantity of offspring; maturing too late is a bad idea because earlier maturation shortens generation time and reduces the risk of dying before maturing; and the quantitative effects of age at maturation on fecundity and on offspring quality (juvenile mortality) differ, both being important.

Phenotypic plasticity for maturation in humans

Do humans vary in age at maturity in a manner consistent with evolutionary expectation? To answer that question, we first have to extend the conceptual framework to include plastic responses to variation in growth rates and physiological condition. A comparison of the life histories of women living in the industrial revolution in the nineteenth century in England and

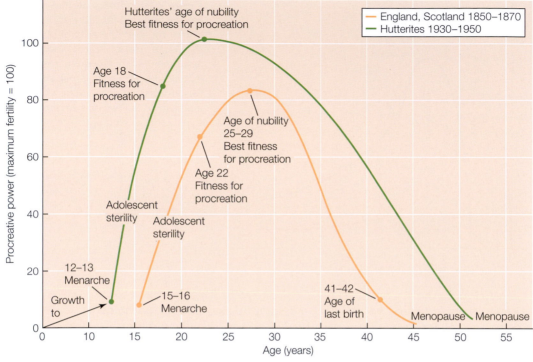

Figure 2.11 Women on twentieth-century Hutterite farms in the United States were better nourished, grew faster, and were younger and larger at maturity than women in nineteenth-century England and Scotland. (After Frisch 1978.)

Scotland with those of women living in Hutterite farming communities in the twentieth century in the United States (Figure 2.11) helps pose the question more precisely: does human age at maturity change to keep fitness as high as possible when faced with stress? If we assume that growth slows or stops with maturation, that larger females have more offspring (a small effect in humans), and that offspring mortality decreases as mother's age at maturity increases (a large effect in humans)—all in a manner consistent with observation—we can predict optimal plastic responses for the human maturation event (Figure 2.12).

The analysis in Figure 2.12, which predicts a 4- to 5-year decrease in age at maturation from the nineteenth century to the twentieth century, agrees fairly well with the observations in Figure 2.11. More important, however, is the way that the figure distinguishes between nurture and nature, the plastic developmental reaction to environmental change and the genetic evolution of that reaction. The upper (blue) curve shows the optimal plastic response to a change in the environment; the maturation event slides up the reaction norm to the left. No evolution is required for that change; it is a developmental response, not a genetic response. However, its shape and position evolved in the past.

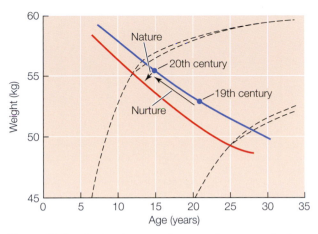

Figure 2.12 Optimal maturation reaction norms for human females. The dashed lines are growth curves. (After Stearns and Koella 1986.)

Will evolution now change the response? That is what the lower (red) curve in Figure 2.12 depicts. When human populations moved through the demographic and epidemiological transition, mortality rates fell, particularly infant mortality rates, as clean water supplies, vaccines, better nutrition, and antibiotics decreased the impact of infectious disease. If that decrease in mortality remains in effect long enough for an evolutionary response with a genetic basis to occur, the entire reaction norm will shift down and to the left, as depicted in the lower curve, resulting in a further decrease in age at maturation of about 6 months and a modest decrease in size. Thus, Figure 2.12 distinguishes between the effect of nature—the genetic evolution of the shape and position of the reaction to the environment—and the effect of nurture—the development of a particular individual in a particular environment—here seen as one point on the line describing the reaction.

Is there any evidence that recent and contemporary human populations are evolving in the direction predicted above? Analysis of the direction of selection on age and size at maturity in populations in Australia, Finland, and the United States all found that selection is currently operating to decrease age at maturity in human females (Stearns et al. 2010), and a study in Canada both confirmed that response and suggested that it had a genetic basis (Milot et al. 2011). Why might that be happening?

The major trade-off affecting the maturation event in humans is the one between infant mortality and age of mother; it is a strong relationship in humans (Figure 2.13). Earlier maturation is evolving because modern hygiene, nutrition, public health, and medicine have decreased infant mortality rates, shifting the curve in Figure 2.13 to the left. Mothers are being selected to have their first baby earlier in life because improvements in health care allow them to do so successfully. They get the benefits of earlier reproduction without having to pay the costs that previously existed. This change has happened because cultural evolution has started to interact

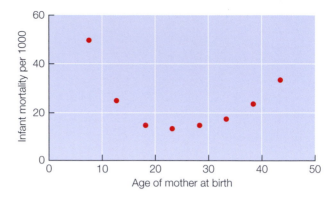

Figure 2.13 Infant mortality rises rapidly as the age of the mother at birth decreases. (After Stearns 1992.)

with biological evolution; here the important cultural change was in health care. Decisions made by physicians and public health workers have started to shape the evolution of their patients as well as their health.

This analysis of age and size at maturity makes several points. What evolves is often not some fixed value of a trait but, rather, the rules that govern the plastic response of that trait to a range of environments. That response, called a reaction norm, is a useful intellectual tool whose properties are further developed below. Reaction norms depict the interaction of nature with nurture, of genes with environments, as mediated by development. In this case, the environmental change consists of cultural improvements in public health and medical care that change selection and alter the course of biological evolution. One consequence is increased conflict between biology, which pushes for earlier maturation, and culture, which now pushes many women to have their first child later.

■ SUMMARY

Natural selection shapes life histories from birth to death. Changes in life history events can increase or decrease reproductive success, or fitness. Selection shapes them to maximize fitness under the constraints of trade-offs, which connect traits via genetics, development, morphology, and physiology. When traits are connected by trade-offs, changes in one trait that increase fitness are linked to changes in other traits that decrease fitness.

Plasticity and reaction norms

Just as age and size at maturity have a plastic response to environmental variation, so do many other traits. As individuals develop, they accumulate a history of environmental encounters that shape the expression of those traits: their genotypes interact with the environment to produce a phenotype. At any given moment in time, a patient is a point on the reaction norm of one trait and the reaction surface of many traits. The rules governing such developmental interactions with environmental variation have evolved and determine the responses produced.

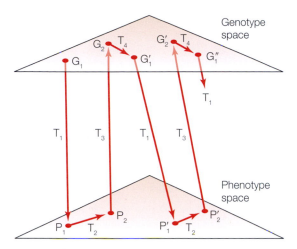

Figure 2.14   Evolution proceeds through a regular generational cycle in which information in genotypes alternates with material in phenotypes. Reaction norms are produced by the developmental interactions of genotypes with environments summarized here as T_1. (After Lewontin 1974.)

Let's now situate those developmental responses in the evolutionary process. As evolution proceeds from generation to generation, there is a regular alternation between the information stored in genotypes and the matter organized in phenotypes (Figure 2.14). Starting from the first set of genotypes (G_1), the first transformation (T_1) maps the information in genotypes into the material in phenotypes. T_1 is development; that is where reaction norms come in. The second transformation (T_2) determines which organisms survive and reproduce; that is where ecology and behavior play their role in natural selection. The third transformation (T_3) involves mating, reproduction, and some genetics; it determines the array of gametes that will form the next generation. The fourth transformation (T_4) involves genetics and reproduction; it determines how the array of haploid gametes is transformed into diploid zygotes, forming a new set of genotypes that complete one cycle and start the next. In every generation, development, ecology, behavior, and genetics each play important roles. We now focus on the first step in the repeating cycle of transformations, development.

Reaction norm basics

A reaction norm expresses the range of trait values that can be elicited from a single genotype by environmental variation. In many insects and amphibians, size at maturity decreases as temperature increases. One genotype might react as in Figure 2.15A; the reactions of an entire population of genotypes would be described by a bundle of reaction norms, each belonging to a different genotype as in Figure 2.15B. Reaction norms can be measured directly whenever organisms can be cloned so that the same genotype can be reared in different environments. In humans, that is the case with monozygotic (identical) twins, such as Otto and Ewald Spitz.

(A) One reaction norm (B) A bundle of reaction norms

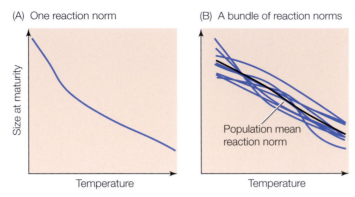

Population mean
reaction norm

Figure 2.15 Reaction norms are properties of genotypes. Individuals are points of a reaction surface. Populations are bundles of reaction norms. (After Stearns and Hoekstra 2005.)

Starting at age 18, Otto ran distance, and Ewald lifted weights. In Figure 2.16, they are 22 years old. Here the reaction norm describes the difference in two bodies produced by the same genotype interacting with different exercise regimes.

The canalization and plasticity of traits are not opposites. Canalization can refer to the buffering of a fixed trait or to the buffering of an entire reaction norm as can be seen in Figure 2.17, which contrasts a bundle of more canalized reaction norms with a bundle of less canalized reaction

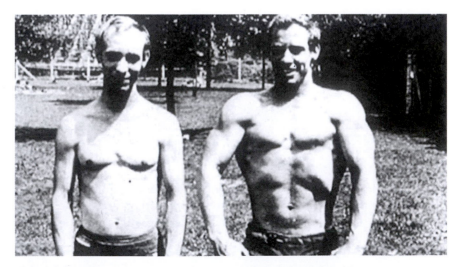

Figure 2.16 Reaction norms in humans can be measured on identical twins. Otto Spitz, on the left, started running distance at age 18. Ewald Spitz, on the right, began lifting weights at the same age. Here, they are 22 years old. (From Kono 2001, courtesy of Tommy Kono.)

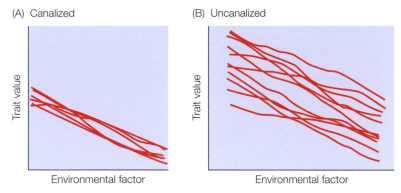

Figure 2.17 Canalization and plasticity are not opposites; plastic responses can be canalized. (After Stearns and Hoekstra 2005.)

norms. In both Figure 2.17A and Figure 2.17B, the traits are plastic: they are sensitive to the environment, as can be seen by the consistently negative slope of the reaction norms. The genetic variation in trait values in any given environment is much greater in Figure 2.17B than in Figure 2.17A, however, as indicated by the greater scatter among the lines, and it is that contrast in variation that leads us to call Figure 2.17A canalized relative to Figure 2.17B.

Some traits are plastic; others are not. Thus, organisms are mosaics of traits with different patterns of plasticity. There is very little genetic variation for number of digits in humans; the vast majority of individuals have five per limb, and, as depicted in Figure 2.18A, that number does not vary with the environment. As in many other organisms, however, human fecundity is plastic. It is reduced when food is limited, and genotypes vary in how their

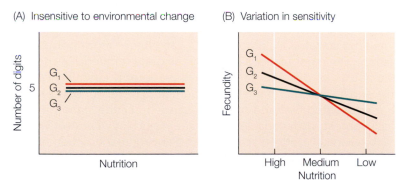

Figure 2.18 In (A), the reaction norms of the three genotypes are separated for illustration; they all describe five digits that do not change with the environment. In (B), the different slopes of the reaction norms for genotypes G_1, G_2, and G_3 show how those genotypes vary in their sensitivity to environmental change. G_1 is quite sensitive, G_3 is fairly insensitive, and G_2 is intermediate. (After Stearns and Hoekstra 2005.)

fecundity responds to food limitation. The slopes of the reaction norms measure variation in sensitivity to the environment: the slope is zero for digits (see Figure 2.18A), and it is negative and variable for fecundity (see Figure 2.18B).

Figure 2.18B depicts an important general situation. Whenever reaction norms cross, there are strong genotype × environment interactions. Unless they are well understood, they can lead to unexpected outcomes. For example, in Figure 2.18B, selection for greater fecundity in the high nutrition environment would cause genotype G_1 to increase in frequency. That would lead to a decrease in fecundity in the population when it is exposed to low nutrition, the opposite of the response in the environment in which the trait was selected.

Genetic correlations among traits are plastic

One way to express the connections among traits that we call trade-offs is by measuring their genetic correlations. If the traits are plastic and if genotypes vary in their sensitivity to the environment, the genetic correlations among the traits will themselves be plastic. When that is the case, the genetic correlation between two traits can actually change from positive to negative across environments. In some environments, selection that increases one trait will cause a correlated increase in the other trait. In other environments, selection that increases one trait will cause a correlated decrease in the other trait. Depending on the relationship of the traits to fitness, such an environmental change may cause the expression or the disappearance of a trade-off.

This situation is not just a theoretical possibility. The genetic correlation between body length and larval period in spadefoot toads switches from negative in ponds of short duration to positive in ponds of long duration (Newman 1988), and the genetic correlation between weight and age at eclosion in fruit flies switches from positive among individuals that are well fed and growing rapidly to negative among individuals that are poorly fed and growing slowly (Gebhardt and Stearns 1988).

Thus, genetic correlations among traits can change from population to population, within populations over time as gene frequencies change, during the course of development (e.g., as measured in young vs. old organisms), and from environment to environment. The responses to selection that are mediated by trade-offs can change in all the same ways. In the medical context, such considerations are important whenever we are dealing with mismatch between patients and environments, whether that is caused by the change from Neolithic to postindustrial environments or by the change in environments encountered by immigrants.

The medical significance of phenotypic plasticity

Although the evolution of reaction norms and the expression of trade-offs as genetic correlations are both basic features of the biology of patients, the immediate medical significance of phenotypic plasticity is probably most apparent in the developmental origins of health and disease (DOHaD). Variation in early-life experience has many consequences, some of which we will explore in more detail below. They include that thin infants are at increased risk of cardiovascular disease, obesity, and diabetes in late life; that infants born by cesarean section (C-section) are at increased risk of asthma, allergies, and obesity; and that children who have more antibiotic treatments before age 2 are at greater risk of obesity and allergies. These consequences are all reaction norms—that is, responses of genotypes to variation in the environment—and they vary among individuals.

It is also important to note that the phenotypic response to environmental change can be reversible (e.g., body weight, muscle mass) or irreversible (e.g., body height); it can be confined to specific developmental stages or produced at any stage (e.g., tanning, physiological responses to altitude); and it can be continuous (e.g., age at maturation) or discontinuous (e.g., alternative morphs in water fleas and social insect castes). The cases referred to under DOHaD concern traits whose change is either irreversible or hard to reverse, locking the organism into a trajectory that adapts it better to the short-term environment but causes problems later in life.

■ SUMMARY

The expression of many traits is sensitive to the environment. The concept of a reaction norm is useful when analyzing how genes and environments interact to produce phenotypes. Reaction norms are properties of genotypes that in humans are most clearly visualized as the differences that develop between identical twins that encounter different environments. Important medical conditions are produced by reaction norms when environments encountered early in life affect the health of adults.

Trade-offs

The trade-off concept was introduced in Chapter 1 and has already been used in several explanations. It is so important that we develop it here in detail.

Patients are bundles of trade-offs that were assembled by a process of evolutionary tinkering with whatever variation was available that worked at that moment. They are not machines designed by an engineer made of replaceable parts that can be exchanged without consequences.

For traits to be involved in a trade-off, two connections must be established. The first is the connection between the traits; it could be positive or negative. The second is the connection between each trait and fitness; it could also be positive or negative. A trade-off—a benefit connected to a cost—only exists when a change in one trait that increases fitness is associated with a change in another trait that decreases fitness.

The connections among traits that lead to trade-offs can be viewed from several perspectives: genetic, developmental, morphological, ecological,

Figure 2.19 The Y model of physiological allocation. (After de Jong and van Noordwijk 1992.)

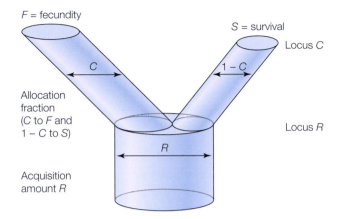

behavioral, or physiological. The physiological perspective was one of the first to be developed.

Trade-offs as energy allocations

We can view the resources acquired by an organism as a pipe with a diameter R (Figure 2.19). If those resources are allocated between only two functions—for example, fecundity (F) and survival (S)—we can visualize the allocation as two smaller pipes defined by allocation fractions that sum to one: fraction C going to fecundity and fraction $1 - C$ going to survival (de Jong and van Noordwijk 1992). To model such a situation genetically, we can assign variation in one gene, locus C, to the allocation fraction and assign variation in another, locus R, to acquisition efficiency. Making those assignments helps us connect physiology to genetics.

The Y allocation model is a simplification with a useful insight: whether we should expect a positive or a negative relationship between the two traits involved in a trade-off depends on which varies among individuals more, acquisition efficiency or allocation fraction. If the variation is mostly in the efficiency with which resources are acquired (Figure 2.20A), the correlation between the traits will be positive: rich people can have both big houses and fancy cars. If the variation is mostly in the fraction allocated to each function (Figure 2.20B), the relationship will be between paying the mortgage and buying a car, and people can have either a fancy house or a fancy car but not both. Such insights help explain why trade-offs often appear under nutritional stress and disappear in well-fed populations.

Where do trade-offs occur, and what causes them?

Trade-offs do occur among physiological functions within the body of a single organism, but we can also usefully see their causes as having other sources. Their evolutionary significance lies in the constraints placed on the simultaneous change of two or more traits, an issue framed at the level of a population experiencing change over generations (cf. Figure 2.14). In kestrels, the clutch size of parents trades off with the age at maturity and

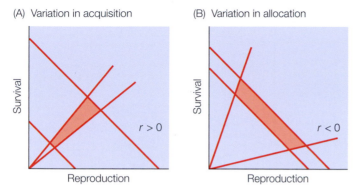

Figure 2.20   The Y allocation model shows that changing the amount of variation in acquisition (A) and allocation (B) of resources can change the correlation of two traits. Positive phenotypic correlation between survival and reproduction is indicated by $r > 0$. Negative phenotypic correlation is indicated by $r < 0$. In both cases, there is an underlying trade-off, and the difference at the phenotypic level is decided by whether variation in acquisition is greater than, or less than, variation in allocation. (After van Noordwijk and de Jong 1986.)

the survival of offspring as well as with the survival of parents. In *Drosophila melanogaster*, older females lay eggs that have higher juvenile mortality than do eggs laid by younger females. In red deer, the effect of reproduction on the mortality of the mother is greater if she has a daughter than a son because the son moves away sooner, whereas the daughter stays nearby and competes for food with the mother. In general, trade-offs can occur within any set of interacting individuals and can be mediated by many types of biological processes, including ecology and behavior as well as physiology, development, and genetics. Any process that contributes to the construction of phenotypes can be involved in trade-offs.

Phenotypic correlations exist for many reasons, not all of them genetic. Only connections among traits that have some genetic basis will produce trade-offs with evolutionary consequences. Therefore, the most convincing way to measure an evolutionary trade-off is as a correlated response to artificial selection: select on one trait and see how the other traits respond. This process has been done perhaps more extensively in the fruit fly *D. melanogaster* than in any other organism. Figure 2.21 presents the results of 14 studies done in five laboratories involving roughly 100 person-years of work. The traits analyzed are in boxes. In each study, a trait was selected to increase or decrease. The arrows run from the traits that were selected to the traits in which significant correlated responses were measured. Positive or negative symbols next to the arrows indicate the signs of the correlated responses.

As seen in Figure 2.21, when early fecundity was selected upward, longevity decreased, and when longevity was selected, early fecundity decreased. When late fecundity was selected, longevity increased. When developmental time was selected upward, early fecundity increased, and so

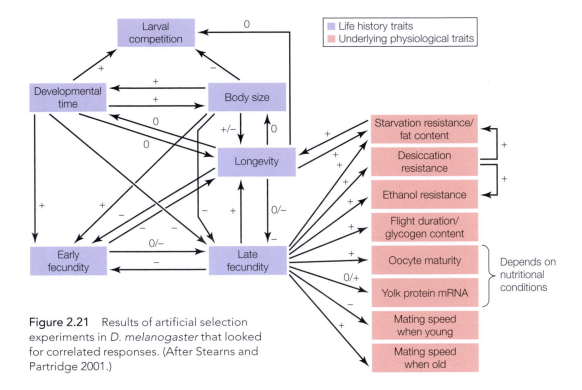

Figure 2.21 Results of artificial selection experiments in *D. melanogaster* that looked for correlated responses. (After Stearns and Partridge 2001.)

forth. These changes in life history traits were accompanied by consistent changes in the underlying physiological traits, shown in boxes to the right.

Responses were not always consistent among laboratories; sometimes no correlated response was observed, and sometimes laboratories reported opposite responses, as indicated by the arrows with 0/–, 0/+, or +/–. To understand why that was happening, flies from several labs were reared according to local practice in each of those several labs. Development time and early fecundity were both found to be sensitive to lab-rearing techniques, and those differences in sensitivity accounted for much of the variation in correlated responses. Flies also grew and reproduced best in the lab environments in which they had been held for generations. They had adapted evolutionarily to their local lab environments (Ackermann et al. 2001).

These studies show that simultaneous evolutionary responses of two or more traits depend on the environments in which they are expressed. These studies are an important background to the attempts to measure costs of reproduction in humans.

Is there a trade-off between reproduction and survival in humans?

The existence of a trade-off between reproduction and survival in humans has been controversial (X. Wang, Byars, and Stearns 2013). Positive, negative, and insignificant phenotypic relationships between completed family size and life span have all been found. A genetic pedigree analysis of more than 5100 men and women living between 1658 and 1907 in South Tyrol,

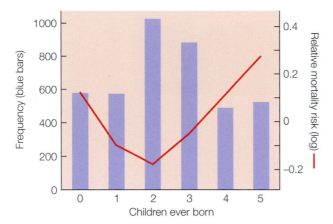

Figure 2.22   The relationship between children-ever-born and mortality risk in women in Framingham, Massachusetts, was U-shaped. (After X. Wang, Byars, and Stearns 2013.)

Italy, found a significant *positive* genetic relationship between completed family size and life span (Gögele et al. 2011); that is not a trade-off. One genome-wide association study of more than 3500 women from Rotterdam found several SNPs associated with completed family size, but none of them also appeared to affect life span (Kuningas et al. 2011); that is also not a trade-off. In the women of the Framingham, Massachusetts, heart study, however, X. Wang, Byars, and Stearns (2013) found that mortality risk declined in those who had one or two children, but then rose markedly in larger families (Figure 2.22); that is a trade-off. The genetic correlation between life span and family size was large, significant, and *negative* (−0.88), but its significance depended on whether education and smoking status were included in the model.

A GWAS found two SNPs marking regions that might be mediating the trade-off, and both regions contained genes associated with cancer risk as well as other functions. Two other studies have also found genes mediating cancer risk where increased risk also increased reproductive performance: the tumor-suppressor gene *p53* (Kang et al. 2009) and the early-onset breast cancer gene *BRCA1* (Smith et al. 2011). Whether there is a general genetic association between reproductive performance and life span mediated by cancer risk remains, at this point, an open issue in need of further study.

Interestingly, mice engineered to have high *p53* activity are cancer resistant but age prematurely (Tyner et al. 2002). Thus, efficient tumor suppression can increase life span due to cancer prevention, but at least in mice it comes at the cost of accelerated aging through other mechanisms, such as cellular senescence or depletion of stem cell pools (Figure 2.23). Elephants, which have multiple copies of *p53*, somehow avoid this trade-off. We do not yet know whether humans resemble mice or elephants more in this respect.

The difficulty in resolving the issue of reproductive costs in humans can be traced to the important role that the environment plays (cf. results

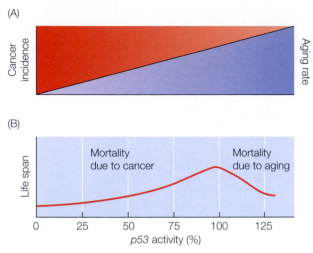

Figure 2.23   (A) Increases in *p53* activity reduce the incidence of cancer but increase the aging rate. Conversely, decreases in *p53* activity increase cancer incidence but may decrease the rate of aging. (B) As a consequence, the influence of *p53* on life span may result from a delicate balance between its antitumor and pro-aging effects, such that too little *p53* increases mortality from cancer, whereas too much *p53* increases mortality from aging. (After Tyner et al. 2002.)

above on spadefoot toads and fruit flies) and to the contrasting effects of variation in acquisition and allocation of resources (cf. Figure 2.20) in mediating the expression of genetic correlations. If such factors could be more precisely controlled, the measurements of this key trade-off would probably not disagree as much.

Hormones and the trade-off between growth, reproduction, and immune function

Because most physiological functions are regulated by the endocrine system, hormones mediate many physiological trade-offs. This is easily seen where hormones regulate resource allocation. For example, insulin controls glucose distribution among tissues. During an immune response, inflammatory cytokines turn down insulin-dependent glucose consumption in skeletal muscle and adipose tissue as well as insulin-mediated suppression of glucose production by the liver. As a result, more glucose becomes available for consumption by the immune system, which takes the priority during an infection. Similarly, inflammatory cytokines can inhibit growth hormone signaling in the liver, resulting in reduced production of IGF-1, a major growth factor controlling body size. Consequently, chronic inflammation during critical growth periods can result in idiopathic low stature syndrome. Whether this condition is caused by resource allocation remains unclear.

In males, testosterone and other androgens regulate the allocation of energy between reproductive and immune functions. During infections, energy is switched from maintaining skeletal muscle mass, red blood cells, and bone density to increasing the immune response, and this reallocation is stronger in males than it is in females. In uninfected males, the normal maintenance of secondary sexual characters by androgens diverts energy away from, or otherwise interferes with, immune function and increases susceptibility to disease. For example, men who received a flu vaccination experienced a drop in testosterone levels for the two weeks following vaccination (Muehlenbein and Bribiescas 2005). Additionally, reduced libido, along with anorexia, fever, and fatigue, are typical components of male responses to acute infections.

In females, leptin, a protein hormone, has a more important role in mediating trade-offs between reproduction and survival. Leptin is secreted by fat cells and regulates energy intake and expenditure through its effects on appetite and metabolism; its concentration in the blood reflects overall fat stores in the body. Leptin stimulates the part of the hypothalamus known as the satiety center, where it signals to the brain that the body has eaten enough. There it inhibits neurons that stimulate eating and stimulates neurons that inhibit eating (Ellison 2009).

When women experience caloric restraint, some reproductive traits respond sensitively, and some do not. Ovarian function, duration of gestation, and final birth weight are sensitive to energy balance, but both the volume and the caloric content of the milk with which the mother supplies energy to the infant are much less sensitive. Pregnant and nursing mothers are strongly stressed by poor nutrition, which causes early births of underweight infants and increases the interbirth intervals of their mothers, but a starving mother will continue to produce milk for her infant until her energy reserves are completely depleted.

Do trade-offs evolve? The case of compensatory mutations

Can evolution alter the strength of an interaction between two traits? We have some evidence from experiments on compensatory evolution in bacteria. Resistance to antibiotics is a benefit to bacteria when they are in the presence of antibiotics, but the ability to metabolize antibiotics imposes a cost on bacteria when there are no antibiotics in the environment. Thus, in the absence of antibiotics, resistant strains would lose in competition with sensitive strains if their trade-off did not evolve. However, it does evolve. Six months of exposure to antibiotics can eliminate the costs of resistance, because compensatory mutations are selected that reduce that cost. When genes for resistance whose genetic backgrounds have been shaped by compensatory mutations no longer experience reduced fitness in the absence of the antibiotic, they are close to neutral in that environment, where they can then persist for a long time (Levin, Perrot, and Walker 2000). In this case, evolution acts rapidly to eliminate a trade-off by reducing costs. We can expect that evolution will act to reduce costs whenever they are regularly encountered and reduction is possible. The remaining trade-offs that we see are those that could not be so altered.

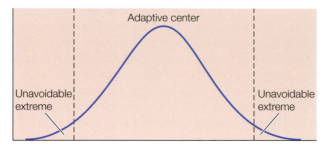

Figure 2.24 Any complex process built from components that vary inevitably produces a distribution of outcomes with a small percentage of extremes.

Deviations from "normal" are unavoidable and often have costs

Any complex process or structure influenced by several to many genes and several to many environmental factors, each of which may vary, will be expressed as a broad range of phenotypes, roughly as a bell-shaped curve (Figure 2.24). Evolution may select for the benefits of performance near the mean, but it has to pay the cost of whatever gets expressed in the tails of the curve. Those costs can be reduced, but they usually cannot be eliminated because both genetic and developmental networks connect the trait states expressed in the tails of this focal distribution to trait states closer to the optimum in the distributions of other traits. Trade-offs are pervasive and frequently multidimensional. The tails of the distribution shown in Figure 2.24 represent costs paid at the population level for benefits realized close to the mean.

Evolutionary geneticists are familiar with one process that contributes to generating the pattern in Figure 2.24. That process is genetic recombination; they call the costs of the extremes that it produces recombinational load. In general, that genetic process combines with variation generated during development, both as developmental noise and as developmental interactions with varying environments, to produce the pattern depicted.

■ SUMMARY

Trade-offs are ubiquitous: they are the default condition. Any trait on which you are focusing is probably involved in a trade-off. Among the traits often encountered in patients are many in which a change that improves one condition worsens another, including the negative side effects of many drugs, the trade-off between androgens and disease resistance, and the trade-off between reproduction and survival. That last trade-off shapes the evolution of aging.

Aging

Aging is defined in evolutionary biology as a decrease with age in the intrinsic ability to survive and reproduce. Aging and life span are not direct objects of selection; they are by-products of selection for reproductive performance. The evolution of aging explains our susceptibility to degenerative disease and why we must grow old and die. All organisms with asymmetric reproduction must age, and all known organisms reproduce asymmetrically. Those are strong, sobering claims. We will see if they stand up to scrutiny, but we first need some background.

Extrinsic and intrinsic mortality

We need to distinguish intrinsic from extrinsic mortality. Extrinsic mortality is caused by factors in the environment, including predation, infection, starvation, dehydration, and freezing. Extrinsic mortality can be nearly eliminated for animals in zoos and humans in modern environments. The remaining causes of death are then intrinsic and a major part of what we think of as aging. The intrinsic causes of human mortality include cancer and metabolic, neurodegenerative, and cardiovascular diseases, all of which we encountered much less frequently in the past when we died earlier for other reasons because extrinsic mortality was greater.

Whereas intrinsic mortality is part of aging, extrinsic mortality affects the evolution of the rate at which we age through its impact on the entire life history, which is shaped by the level of extrinsic mortality to optimize reproductive success. High extrinsic mortality causes the evolution of earlier maturation, increased fecundity, less investment in maintenance, and, consequently, shorter intrinsic life span. Low extrinsic mortality causes the evolution of later maturation, lower fecundity, greater investment in maintenance, and longer intrinsic life span.

The major causes of extrinsic mortality have twice changed dramatically in recent human history, first with the transition from hunter-gatherer to agriculture, and second with the industrial revolution and demographic transition. Our life histories have not yet had time to evolve a new equilibrium that matches current levels of extrinsic mortality, which are much lower now than they have been for most of our evolutionary history.

The key assumptions of the evolutionary theory of aging

The evolutionary theory of aging is based on two important effects. First, as noted by Medawar (1952), the probability that an organism will contribute to the next generation progressively decreases with age once it starts reproducing. That would be true even in a species that did not age because there are always events that kill individuals for reasons that are not related to age. Evolution is largely a numbers game. Selection acts more effectively on events that occur more frequently (reproduction by the young) than it does on those that happen less frequently (survival in the old).

Second, building on Medawar's insight, Williams (1957) pointed out that a new mutation that improved reproductive performance early in life at the cost of reduced survival later in life would be selected whenever the benefit of the increase in early reproductive performance outweighed the cost to fitness of the decrease in later survival. We call such effects antagonistic pleiotropy: they are pleiotropic because the mutation is affecting at least two traits, and they are antagonistic because its effects on one trait increase fitness while its effects on the other trait decrease fitness.

Note that selection is operating on a gene that improves the reproductive performance of the organism so that the gene increases in frequency in the population more rapidly. Once the gene is successfully transmitted to the next generation, the organism that carried it is less important than before. When genes with such effects accumulate in the population, they create trade-offs between reproduction and survival.

The causes of aging

The conditions that cause evolutionary change in life span follow from Medawar's insight about the contributions to the next generation made by organisms of different ages. Longer life is selected by lower extrinsic adult mortality rates or higher extrinsic juvenile mortality rates; they increase the relative contribution to the next generation of the adults while decreasing that of juveniles. In contrast, when adult mortality rates increase or juvenile mortality rates decrease, organisms will evolve to age more rapidly and will then have shorter intrinsic life spans; after all, why should they invest in maintaining a body that will soon be dead anyway for reasons that cannot be controlled? It would be better for them to reproduce while they can.

Aging is thus a by-product of selection for reproductive performance. It arises through the evolutionary accumulation of mutations that have positive or neutral effects on fitness components early in life and negative effects on fitness components late in life. Mutations that have positive effects at all ages will be rapidly fixed, those that have negative effects at all ages will be rapidly eliminated, and genes with a mixture of positive and negative effects will accumulate to create trade-offs.

Because organisms are bundles of trade-offs, because aging is a by-product of a reproductive performance as shaped by those trade-offs, and because patients differ both genetically and developmentally, we expect aging to have many proximal causes that differ among individuals. If we fix one problem, another will emerge, and the problem we try to fix in one patient will differ from the problem we try to fix in another patient. When we reduced mortality caused by infectious diseases by supplying clean water, vaccines, and antibiotics, cardiovascular disease emerged as a major killer. When we then improved the prevention and treatment of cardiovascular disease, mortality from cancer and dementia increased. It is not clear that this process has an end. The evolutionary theory of aging suggests that it does not. The implication is that immortality is impossible to achieve.

Why the body is disposable

If we recast the trade-off between reproduction and survival as a trade-off between reproduction and repair (in which we include immune function as well as intracellular maintenance), the logic of Medawar and Williams can be translated into the language of cellular functions (Kirkwood and Holliday 1979) as displayed in Figure 2.25. There are two key messages in

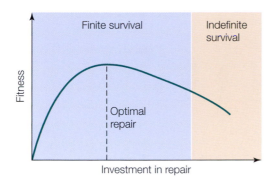

Figure 2.25 The body is disposable because optimal fitness is achieved at a level of repair lower than that needed for indefinite life. (After Kirkwood and Cremer 1982.)

this figure. First, the relationship between fitness and repair has an intermediate optimum. That makes sense if we think about very low and very high levels of repair. With low repair, the organism dies quickly, with no chance to reproduce. With high repair, so much is invested in keeping the organism intact and functioning that nothing is left over for reproduction. Some intermediate level of repair must be optimal for reproductive performance. Second, the level of repair required for indefinite survival is located somewhere to the right of the optimum. That will always be the case when enough genes with antagonistically pleiotropic effects have accumulated in the genome, a property of evolved biology but not of a priori engineered design. The consequence is that investment in repair will evolve to a level lower than that needed for indefinite life. That is why we grow old and die.

Who should age? Are any organisms potentially immortal?

The biological criteria that might separate inevitably mortal from potentially immortal organisms have been much discussed. Weismann (1882) implied that aging is a property of asymmetric reproduction in which the offspring receives newer parts in better condition, but his idea was forgotten. Williams (1957) thought that aging is a property of sexual organisms with separation of the soma, which is mortal, from the germ line, which is potentially immortal, but we now know that bacteria, which do not have a germ line, also age (Ackermann, Stearns, and Jenal 2003). Partridge and Barton (1993) rediscovered Weismann's criterion, suggesting that aging should be a property of any asymmetrically reproducing organism in which a mother (older) can be distinguished from a daughter (younger). This suggestion appears to be true.

When division is perfectly symmetrical, it is impossible for natural selection to distinguish between the two cells; both are equally intact or equally damaged. Therefore, the reproductive payoff from improving the maintenance of both is equal, and symmetrically reproducing organisms should not evolve aging. In contrast, in asymmetrically reproducing organisms in

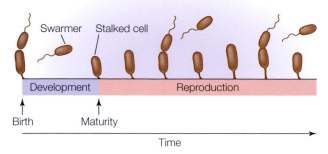

Figure 2.26 *Caulobacter* is an asexual bacterium with asymmetric reproduction. (After Ackermann et al. 2001.)

which a daughter can be distinguished from a mother, aging will evolve when the payoff of maintenance in the older mother becomes smaller than the payoff of maintenance in the younger daughter.

The first bacterium in which aging was demonstrated was *Caulobacter*, an organism that has a sessile adult stage (the stalked cell) and a motile juvenile stage (the swarmer cell) (Figure 2.26). In *Caulobacter*, reproductive performance declines with age because the interval between cell divisions increases (Figure 2.27) (Ackermann, Stearns, and Jenal 2003).

Reproduction in *Caulobacter* is overtly asymmetrical; the stalked and swarmer cells are easily recognized in a microscope. Asymmetry can also be cryptic. *Escherichia coli* has apparently symmetrical division under the microscope, but if the cell poles are labeled so that old can be distinguished from new, daughter cells with new poles can be distinguished from mother cells with old poles. Over generations, the inheritance of new

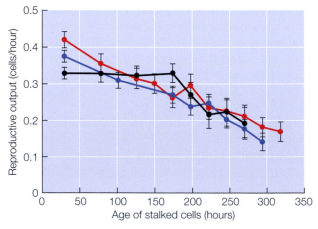

Figure 2.27 In *Caulobacter*, reproductive performance declines with age. The red, black, and blue lines indicate three replicates. (After Ackermann et al. 2003.)

versus old cell poles has cumulative effects. Cell lineages that receive the new poles reproduce more rapidly than cell lineages that receive the old poles, with mixtures of the two having intermediate fitness. As a result, the old lineages die out and are replaced by the new lineages (see Figure 2.1).

Thus, even a bacterium whose reproduction is as apparently symmetrical as *E. coli* actually reproduces asymmetrically, and it ages. If a relatively simple organism with those properties ages, it is likely that all organisms age. Aging remains to be demonstrated in some fungi and a few coelenterates, including *Hydra*, but we do not yet know whether they are exceptions that demonstrate a fundamental principle or exceptions that prove the accepted rule.

As mentioned above, the process illustrated in Figure 2.1 helps account for the maintenance of the germ line since the origin of life. The mechanisms maintaining its function include both molecular repair within germ cells and natural selection among gametes.

What mechanisms mediate aging?

There are many physiological correlates of aging, but those thought to be causes are fewer. Of the causes, two have received particular attention.

The first cause is proton leakage from metabolism. When adenosine triphosphate (ATP) is generated in mitochondria, protons are a by-product. They react with water to form highly reactive peroxides that cause oxidative damage; some leak out of the mitochondria into the cytoplasm, from where they can get into the nucleus and cause mutations. Increases in growth rate and metabolic rate cause energy flow to increase, suggesting that genes mediating energy metabolism could play roles in aging. Two such pathways have been identified: the growth hormone and the insulin/insulin-like growth factor pathways. Both spontaneous and targeted mutations in mice that disrupt these pathways increase life span by up to 70%, and similar effects have been observed in flies and worms. Humans, however, are not model organisms, and mutations in homologous genes, which have been found in one cluster of dwarves living on a Croatian island, result in individuals who remain prepubescent, the size of 7-year-olds, and are chronically cold and sleepy (Austad and Finch 2008). Such issues aside, it is clear that oxidative damage to cells is involved in aging and that repair mechanisms have evolved to limit it.

The second cause is telomere damage. Telomeres are protective caps on the ends of chromosomes that are shortened with each cell division. When they erode away, the next cell divisions fail to replicate the entire genome. Telomere length is maintained by an enzyme, telomerase, which varies among individuals. Centenarians have been observed to have unusually effective variants of telomerase, as do cancer cells with the capacity to divide many times. It is not yet clear whether telomerase is involved in trade-offs with performance earlier in life—for example, with cancer risk—or whether the variants found in populations are essentially neutral because their effects are not registered until after reproduction has been completed.

Could humans be selected to live longer?

There does not appear to be an upper limit on the life span of biological materials if cells can be renewed. The best evidence for that statement is the survival of the germ line, which stretches back in an unbroken sequence of generations to the origin of life about 3.5 billion years ago. Model organisms respond significantly to artificial selection for increased life span, usually with correlated decreases in reproductive performance early in life (cf. Figure 2.21). Some multicellular animals—clams, lobsters, and tortoises—live several hundred years, and some multicellular plants and fungi live considerably longer than that.

Could humans be selected to live longer? That might be arranged if everyone agreed to delay reproduction, first to age 30; then, after life span had increased, to age 40; then to age 50; and so forth. Selection would improve physiological function throughout the lengthening juvenile period; first 30, then 40, and then 50 would be the new 18. The heritability of life span in humans is about 0.25; life span can respond to selection. Would that process go on indefinitely? It would not. An upper limit on life span would be reached, determined by the part of extrinsic mortality that cannot be eliminated, including accidents. At that point, life span would be somewhere between several centuries and perhaps a millennium.

This exercise is interesting, but it is a fantasy that could never be implemented because many people would be so-called defectors from the covenant. They would choose to reproduce earlier and would have correspondingly shorter lives, and their genes would be favored.

■ SUMMARY

Aging evolves in any organism that reproduces asymmetrically, which appears to include most organisms, certainly us. It does so for two reasons: selection intensity declines with age when older mothers can be distinguished from younger daughters, and mutations that improve early performance at the cost of survival cannot be prevented from accumulating in populations. Thus, aging is not directly selected: it is a by-product of selection for reproductive performance. The major assumptions and predictions of the evolutionary theory of aging have been repeatedly tested in many types of organisms and frequently confirmed.

The unusual human life history

The human life history differs from those of our closest living relatives, the chimpanzees and bonobos, in several traits with medical significance (Table 2.8). Our infants are large, fat, and relatively undeveloped. They are weaned early, enabling shorter interbirth intervals. They develop slowly and have an extended childhood that is unique to humans. Both traits imply help with reproduction, a social support network. Humans delay maturity and usually give birth to one offspring at a time, occasionally to two. Human females experience menopause, resulting in an extended

TABLE 2.8 The human life history evolved recently

	Humans (Homo sapiens)	Bonobos (Pongo pygmaeus)	Chimpanzees (Pan troglodytes)	Gorillas (Gorilla gorilla)
Female weight (kg)	40.1	37.0	31.1	93.0
Male weight (kg)	47.9	42.5	41.6	160.0
Gestation length (days)	267	260	228	256
Birth weight (kg)	**3.30**	1.73	1.76	2.11
Age at weaning (years)	**2.0**	3.0	4.0	4.3
Female age at first breeding (years)	**19.3**	10.7	11.5	9.9
Average maximum life span (years)	**70+**	50	45	39
Interbirth interval (years)	**3.5**	2.8	5.0	4.0

Source: After Stearns, Allal, and Mace 2008, Table 3.1, page 48.

period of postreproductive survival. Both sexes have a relatively long life. Let's now take a tour of the human life history. We start before conception.

Oocytic atresia and selective spontaneous abortions

The biology of gametes and oocytes suggests that quality control has long been an evolutionary concern. At 3 months from conception, a human female fetus develops ovaries with about 7 million oocytes. They are then progressively eliminated in a process called oocytic atresia; at birth, she has about 1 million; by menarche, she has about 1500. Why make millions of oocytes and then kill most of them? Some evidence suggests that the genetic quality of the oocytes increases during this process: it is the defective oocytes that are being preferentially discarded.

Quality control continues after conception. As many as 70% of concepti may be aborted in the first menses following fertilization. Many of them have abnormal chromosome numbers and early developmental defects. Evidence from Hutterite women with recurrent spontaneous abortions suggests that concepti with paternal major histocompatibility complex (MHC) alleles identical to the maternal copies are discarded either to avoid inbreeding depression or to eliminate offspring whose deficient immune function would not allow them to survive childhood diseases (Ober et al. 1992). Ultrasound, now a standard obstetric procedure, has revealed that about 70% of conceptions that start as twins end as singleton births. Evolution has designed the human female reproductive tract to function as a quality-control device.

Size at birth: Where many trade-offs start

At birth, human babies range from about 1 kg to 4.5 kg and from about 40 cm to 57 cm. They have larger brains and are fatter than babies born to similar-sized chimpanzees or bonobos, and that extra fat is thought to be a savings account that buffers the fat-needy and long-developing human brain against nutrient shortages that may occur early in life. Human brains continue to grow and differentiate for about 7 years after birth; our closest primate relatives complete brain development much earlier than that. Relative to adult brain size, human babies should have larger brains at birth achieved through longer gestation, but development is constrained by the diameter of a birth canal shaped by upright posture and bipedal locomotion. The resulting compromise produces babies that on average are about 0.5 kg less than the optimal birth weight for infant survival. Parent-offspring conflict may also have a role in that result. Both very low and very high birth weights are associated with poorer survival, health, growth, and mental development than are average birth weights. Events happening in this early period of life have many consequences later in life that we discuss below.

Growth patterns

Human growth is target-seeking, self-stabilizing, and divided into five periods: infancy (birth to weaning at about 2 years), childhood (weaning to full brain growth at about 7 years), juvenility (from the cessation of brain growth to the beginning of puberty at about 10 years in girls and 12 years in boys), adolescence (from the beginning of puberty to full sexual maturity at about 14 years in girls and 16 years in boys), and adulthood (sometimes catch-up growth can continue into the mid-20s). Childhood is not only a period during which the brain develops to an unusual extent; it may also be an adaptation that enables shorter interbirth intervals because, after weaning, children can do some foraging on their own and get supplemental food from members of their extended family.

Females and males have strikingly different growth patterns (Figure 2.28). Although they share a uniquely large adolescent growth spurt, that growth spurt occurs in girls 3 to 4 years before it does in boys. Moreover, the timing

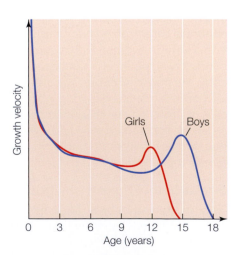

Figure 2.28 Human growth has a unique adolescent growth spurt that differs strikingly in timing between girls and boys. (After Tanner and Davies 1985.)

of expression of key reproductive traits is reversed in the two sexes. Girls look adult and have menses before they are fertile, which may allow them to learn adult roles before experiencing the risks of pregnancy. Boys, in contrast, produce functional sperm before they complete growth and develop adult appearance, which may allow them to father children without eliciting as much competitive male aggression as they would encounter if they appeared adult. Men are slightly larger than women and are more muscular; women store more fat in thighs, buttocks, and breasts than do men.

Humans are less sexually dimorphic in body size than are chimpanzees or gorillas (see Table 2.8). This difference may indicate that polygyny and competition for mates played a smaller role in our evolutionary history than it did in theirs.

Patterns of reproductive investment

The basic pattern of human reproduction prior to birth control is displayed in the contrast between Emma Darwin, Charles's wife, and Fifi, a female chimpanzee in Jane Goodall's study (our thanks to David Haig for this example). Emma had 10 children in 17 years; 7 survived. Fifi had 9 children in 31 years; 7 survived. Emma's interbirth interval was 1.8 years; Fifi's was 3.9 years. Emma had her first child when she was 31 years old; Fifi had hers when she was 13. Emma had her last child when she was 48 years old and died when she was 88. Fifi had her last child when she was 44 and died when she was 46.

This evolved pattern of delayed maturity followed by multiple births at short intervals is now modified by several factors, some cultural, some biological. Nutrition, stress, lactation, and venereal disease intervene on the biological side; marriage, contraception, and conscious decisions intervene on the cultural side. The mother's fertility trades off with infant survival because shorter interbirth intervals endanger offspring at both ends and larger families increase the risk of mortality in 2- to 4-year-old children.

That both biology and culture affect contemporary human reproductive decisions does not mean that our reproductive pattern has stopped evolving. Biology and culture combine to affect offspring survival rates as mediated by disease, nutrition, and the presence or absence of mates and relatives who can help with child care. When infant and juvenile mortality are decreased, so are the costs of earlier maturation, and natural selection then favors earlier age at first birth. Whether that actually occurs depends on both biology and culture, which may be in conflict.

Why don't humans have more children in each reproductive event? Only about 4% of natural births are twins. Studies of preindustrial societies have shown that twinning can increase lifetime reproductive success in some environments but not in others (Lummaa, Jokela, and Haukioja 2001). In the eighteenth and nineteenth centuries, the Finnish archipelago had a richer and more stable food supply than did the Finnish mainland. Twins born in the archipelago had better survival than did twins born on the mainland and on average produced more than 1 adult, whereas twins on the mainland produced on average only 0.7 adult. Twin sisters were more successful; twin brothers were less successful. Mothers of twins had higher mortality after age 65 because they were more susceptible to tuberculosis. Such studies reveal trade-offs between offspring number and both offspring survival and parental survival.

In food-stressed populations, twinning is maladaptive. In some hunter-gatherer populations, one twin is killed at birth, increasing the likelihood that the other will survive. Evolution may have anticipated this measure because, as noted above, about 70% of conceptions that start as twins end as singletons; one of the two embryos is reabsorbed early in pregnancy.

Life span and aging

That we have longer intrinsic life spans than chimpanzees and bonobos is, in part, the evolutionary result of the lower risk of mortality caused by our superior social organization and group defense against predators. In addition, humans can arrange, through parental care and group interactions, intergenerational transfers of resources and care that increase the contributions to reproductive success of older individuals, whether they are mothers and fathers or grandmothers and grandfathers (Lee 2003b). There is also a third effect, this one between the sexes. Men can reproduce later in life than women because they can have a series of partners in sequence. Selection to prolong male life may then indirectly prolong female life because many of the same genes and regulatory networks are involved in maintenance in both sexes.

Menopause

Chimpanzees do not have menopause; healthy females continue to bear offspring until they die. In contrast, in humans menopause is normal. A trait that evolved in our lineage since we last shared ancestors with chimpanzees, it is puzzling because at first sight it would appear that menopausal females are forgoing significant opportunities to increase their reproductive success. There are at least four explanations for its evolution (see also Chapter 7).

The first is about mothers, who stop reproducing because the risk of dying in childbirth increases with age and they should survive long enough to rear their last child. This explanation is the mother hypothesis (Williams 1957).

The second is about grandmothers, who can better help their daughters rear their grandchildren if they stop having children themselves (Hawkes et al. 1998). This explanation is the grandmother hypothesis.

The third is about reproductive conflict. Mothers-in-law are more related to their own offspring (0.5) than to those of their daughters-in-law (0.25), whereas daughters-in-law are unrelated to their mother-in-law's offspring (0.0) and more related to their own (0.5). Thus, daughters-in-law are under stronger selection to win conflicts over resources used for reproduction. This effect of asymmetrical relationship could select against continued reproduction in older women if they compete with their daughters-in-law for resources. This explanation is the daughter-in-law hypothesis.

The fourth is about offspring quality. It sees menopause as a by-product of selection for efficient quality control early in life. Menopause occurs when the supply of oocytes is exhausted by atresia. Under this hypothesis, the improvement in offspring quality achieved by stringent selection of high-quality gametes early in life more than compensates for lost reproduction late in life. The standard logic for the evolution of aging applies. This explanation is the quality-control hypothesis.

Evidence on these hypotheses is mixed. The mother hypothesis only works by itself if the increase with age in the risk of dying in childbirth is implausibly large. There is some evidence for and some evidence against the grandmother hypothesis as a single explanation. Reproductive conflicts between mothers-in-law and daughters-in-law appear to be real in some circumstances, but how much they could contribute to a multifactor explanation is not yet clear: there is some question about the size of the effect. There is as yet no convincing evidence for or against the quality-control hypothesis; we do not know how much the fitness of an offspring who has passed through the filter of atresia exceeds that of an offspring from a mother whose filter is defective. The hypotheses are not mutually exclusive, and some studies have found effects that in combination are sufficient to explain menopause.

■ SUMMARY

Like all life histories, the human life history is an evolutionary solution to a set of ecological problems. Our short interbirth interval is striking and possible because of social help, both from mates and from other family and group members. We are large and notably fat at birth, and that extra fat is one reason that our brains are able to continue to grow for another 7 years. There are significant trade-offs between the intervals at which mothers give birth and the survival rates of their offspring, between physical vigor and disease resistance in males, and between reproduction and maternal survival. Menopause is a special feature of human life histories that probably evolved for a combination of reasons that include effects mediated by mothers, by grandmothers, by daughters-in-law, and by quality control of oocytes.

Developmental origins of health and disease

Conditions in utero, in infancy, and in childhood affect the risk of cardio-vascular disease, diabetes, obesity, and mental illness much later in life. These are plastic responses to environmental variation in part shaped by past selection. They exist for two general reasons. First, benefits realized early in life are associated with costs paid much later at ages to which many might not have previously survived: the net benefit is then positive. Second, the expression of these traits may be constrained to be irreversible. Both characteristics apply to some but not to all traits. The expression of some traits is adaptively regulated so that they are expressed only when beneficial. The relationship of some traits to fitness is such that they are adaptive both in the young and in the old when a particular environment induces them in the young. The expression of all these traits may either be in the form of a continuous reaction norm or as discontinuous phenotypes (polyphenisms).

Many noninfectious diseases originate early in development. If the developing organism does not encounter its normal symbionts, their evolved interactions with the immune system do not take place, leading to abnormal immune responses, including allergies and autoimmune diseases. Early events also contribute to cancer because the somatic mutations that lead to cancer are most influential if they occur early in development when cell division is rapid and downstream influence is greatest.

We deal with abnormal immune responses and cancer later. Here we focus on the metabolic syndrome, where condition at birth affects the probability of type 2 diabetes, obesity, high blood pressure, and cardiovascular disease 50 to 60 years later.

Two early hypotheses: Thrifty genotypes and thrifty phenotypes

The puzzling existence of diabetes has elicited several attempts to explain it. Neel (1962) suggested that genes for insulin resistance were selected by famines in past environments, leading to mismatches in modern ones. This thrifty-genotype explanation, based on genetic determinism, called attention to the problem, but famines do not appear to have been frequent enough in our evolutionary past to generate a selection advantage large enough to compensate for the costs of diabetes at times when nutrition was not limited. Hales and Barker (1992) then suggested that conditions in utero and in infancy set developmental switches that launch patients on trajectories conditioned by early environment. Their thrifty phenotype explanation, based on developmental determinism, tried to explain why the metabolic syndrome is more frequent in adults who were undernourished early in life, but it also had a weakness. Adult environments, which occur decades later, are usually too weakly correlated with infant environments to let infant condition be a good predictor of the issues adults will face. If the induced state triggered by early nutritional stress is an adaptation at all, it is more likely to be a response designed to protect the infant and young child from starvation, with the consequences for adults being costs imposed by the delayed effects of a trade-off.

The Dutch Hunger Winter

The data that initially prompted the thrifty genotype and thrifty phenotype hypotheses came from a dramatic event at the end of World War II. In the winter of 1944–1945, from December to May, the Nazis blocked food shipments to Holland. The resulting famine had long-lasting consequences for the children of mothers who were pregnant at the time. Those consequences differed depending on the period of pregnancy during which the fetus encountered the stress: late, middle, or early (Figure 2.29). If the stress occurred in early gestation, the risk of cardiovascular disease and general

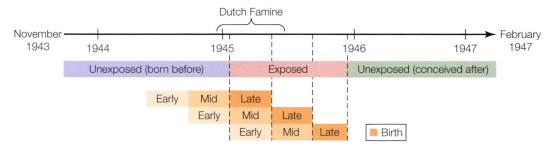

Figure 2.29   The Dutch famine birth cohort encountered peak stress at different periods of pregnancy. (After Roseboom et al. 2001.)

poor health increased in adults. If it occurred in midgestation, the risk of kidney disease associated with diabetes and of lung disease increased in adults. If it occurred in late gestation, the risk of type 2 diabetes increased (Table 2.9).

Conditions in the Netherlands in the winter of 1944–1945 were extreme; people were forced to survive on fewer than 1000 calories per day and sometimes much fewer (Roseboom et al. 2001). Do similar effects occur under more normal conditions?

The effects of pregnancy-related stress under more normal conditions

The discovery that conditions early in life affected health late in life prompted an explosion of research, one of whose aims was to see whether effects elicited by extreme conditions also occurred under more normal circumstances. It appears that they do. For example, Barker and colleagues (1989) studied the relative risks of ischemic heart disease in adult men, breast-fed as infants, as a function of their weights at birth and at 1 year (Figure 2.30).

The average infant to whom risk was normalized at 100 was about 7 lb 12 oz at birth and about 22 lb at 1 year. Deviations downward from those weights were associated with increased risk of heart disease; deviations upward, with

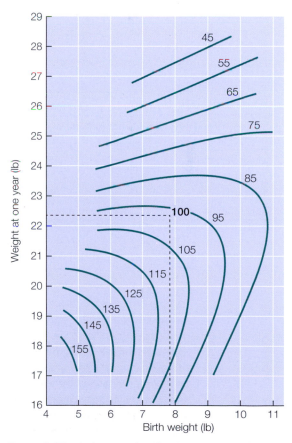

Figure 2.30   Relative risks of ischemic heart disease in men born in Hertfordshire as a function of weight at birth and at 1 year. (After Barker et al. 1989.)

TABLE 2.9 Adult characteristics according to timing of prenatal exposure to the Dutch famine: Mean and standard deviation

Adult characteristics	Exposure to famine						
	Born before	Late gestation	Mid-gestation	Early gestation	Conceived after	All (S.D.)	n
Number	264	140	137	87	284		912
Proportion of men	49%	47%	39%	41%	50%	47%	
Plasma glucose 120 min[a] (mmol/L)	5.7	6.3[b]	6.1	6.1	5.9	6.0 (1.4)	702
Plasma insulin 120 min[a] (pmol/L)	160	200[b]	190	207	181	181 (2.4)	694
LDL/HDL cholesterol[a]	2.91	2.82	2.69	3.26[b]	2.94	2.90 (1.53)	704
Fibrinogen (g/L)	3.02	3.05	3.05	3.21	3.10	3.07 (0.6)	725
Factor VII[a] (% of standard)	128	131	133	117[b]	133	129 (1.4)	725
Body mass index (kg/m^2)[a]	26.7	26.7	26.6	28.1	27.2	27.0 (1.2)	741
Coronary heart disease	3.8%	2.5%	0.9%	8.8%[b]	2.6%	3.3%	736
Microalbuminuria (ACR ≥ 2.5)	8%	7%	12%[b]	9%	4%	7%	724
Systolic blood pressure (mmHg)	126.0	127.4	124.8	123.4	125.1	125.5 (15.5)	739
Diastolic blood pressure (mmHg)	86.2	86.4	84.4	84.8	85.2	85.6 (9.9)	739
Obstructive airways disease	15.5%	15.0%	24.8%[b]	23.0%	17.3%	18.1%	733
General health poor	4.5%	6.4%	3.7%	10.3%[b]	5.3%	5.5%	912

Source: After Roseboom et al. 2001, Table 2, p. 96.

Note: n = number of individuals.

[a]Geometric mean.

[b]p corrected for gender <0.05 compared to unexposed (born before and conceived after the famine).

decreased risk. Analysis of people born in Helsinki between 1934 and 1944 suggested a similar conclusion: both males and females who had greater risk of coronary disease as adults had been at least 0.2 standard deviations below average body mass index from birth to about age 6 (Barker et al. 2005). Evidently, the sensitive period early in life extends past birth through infancy and into childhood. The mechanism may be epigenetic and may involve the methylation of the *RXRA* gene, which mediates the effects of retinoic acid on metabolism (Godfrey et al. 2011).

Is the response adaptive?

Similar responses have been found in other studies of humans and in several model organisms; they appear to be real and conserved in evolution.

Whether they are just responses to the immediate environment, aimed at the immediate survival of infants and children, or are also responses to predicted future environments, aimed at conserving energy under possible future nutrient stress, is not yet settled. In either case, the costs in terms of adult disease are paid much later at ages that contribute less to fitness, like the genetic effects of antagonistic pleiotropy in the evolutionary theory of aging, except that here the effects are developmental, not genetic. Although that observation might make the situation more understandable, it does not explain why the costs must be paid at all. If there is an unavoidable constraint that forces their payment, it has not yet been found.

■ SUMMARY

Some events occurring early in life have consequences for late-life health and disease. Environments experienced by fetuses, infants, and children elicit plastic responses that change the risk of adult disease, including cardiovascular disease, diabetes, and obesity. Although initially observed under the extreme conditions of famine, it is now known that the effects occur within the normal range of conditions experienced by fetuses, infants, and children. The mechanisms mediating the response are likely to be epigenetic, and some epigenetic effects have indeed been found. The evolutionary rationale for the responses observed, however, is not yet clear.

The microbiota and their microbiome

A patient consists of about 10 trillion human cells and 90 trillion cells of bacteria and fungi as well as an unknown but presumably much larger number of viruses, including bacteriophages. This assembly of symbiotic microorganisms (collectively referred to as the microbiota) contains 300 times as many microbial genes (referred to as the microbiome) as human genes. The microbiota share a long coevolutionary history with us. They have come to depend on us, and we have come to depend on them. Symbiotic microorganisms colonize every tissue exposed to the environment, including skin, gastrointestinal tract, urogenital tract, upper airways, and even the placenta. Our microbiota provide us with several essential benefits, including protection from infections, metabolic capacities to digest certain foods and synthesize essential nutrients, and signals that initiate the proper development and function of the immune system. Several modern interventions disrupt the natural development of the microbiota. They include C-sections, infant formula rather than breast milk, diet, antibiotic treatments, and hygiene. By altering the interactions of the microbiota with the immune system, these interventions increase the risk of allergies, asthma, obesity, and autoimmune disease.

Codevelopment: Evidence of coevolution

The gut is the largest immune organ in the body. It is surrounded by lymphoid tissue that defends against any bacteria that might penetrate the gut wall. As shown by experiments on infant rabbits contrasting bacteria-free controls with bacteria-inoculated treatments, the development of this gut-associated lymphoid tissue (GALT) is triggered by signals produced by gut bacteria, one of which is *Bacterioides fragilis* (Maynard et al. 2012). The development of GALT is essential to survival; thus, that the mammalian genome outsources this key function to gut bacteria is strong evidence that our lineage could always rely on their presence. They have been a part of our bodies for many millions of years.

Colonization by symbiotic microbes presents a unique challenge to our immune system: it has to eliminate invasive pathogens, but at the same time it must spare harmless and beneficial symbionts. The difference between the two is not always obvious, and the immune system can and does sometime make mistakes. When that happens, the consequence can be the devastating inflammatory bowel diseases (IBDs), which include Crohn's disease and ulcerative colitis. Because immunological balance is maintained by multiple mechanisms, there are several paths to disease (Izcue, Coombes, and Powrie 2009). For example, both mutations in genes that suppress the immune response—such as the gene encoding the IL-10 receptor (Glocker et al. 2009)—and mutations in genes that initiate it—such as the gene encoding one of the microbial sensors, called NOD2 (Cho 2008)—significantly increase the risk of IBDs.

The maturation of the gut immune system is triggered by postnatal bacterial colonization (Figure 2.31). It stimulates the growth of Peyer's patches, which are lymphoid nodules surrounding the ileum; cryptopatches, which mature into isolated lymphoid follicles; and mesenteric lymph nodes. These elements of GALT all contain populations of dendritic, T, and B cells involved in signaling and defense. In the gut epithelium itself are Paneth cells. They secrete antimicrobial compounds into the gut lumen, joining the major immunoglobulin in gut mucus, IgA, which is produced by B cells (Maynard et al. 2012), and goblet cells, which produce mucin.

Together these cells secrete products that create an inner mucous layer next to the gut epithelium that is nearly sterile and thicker in the large intestine than it is in the small intestine. The gut microbiota live in the outer mucous layer and in the gut lumen. That is the structure of the gut environment in which our immune system "farms" our microbiota.

The microbiota develop across at least two generations. Before the mother becomes pregnant, her microbiota are shaped by her diet, her genes, and the epigenetic programming that reflects her developmental history. While she is pregnant, her microbiota are affected by stress, smoking, nutrition, and placental signals. During vaginal delivery, bacteria from the mother's vagina and gut colonize the infant. That does not happen, at least not to such an extent, when delivery is by C-section. After birth, decisions about antibiotic treatments and breast-feeding versus infant formula affect further colonization and the balance of bacterial species in the infant gut. In early infancy, the microbiota continue to develop under the influence of diet, including potential allergens, infections, and secondhand smoke.

(A) Prenatal

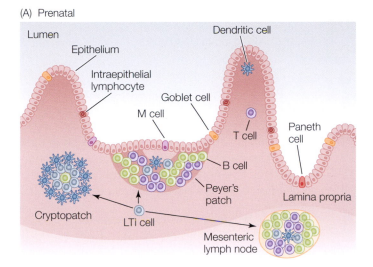

Figure 2.31 Postnatal bacterial colonization transforms the gut environment and stimulates the development of the immune system. (After Maynard et al. 2012.)

(B) Postnatal

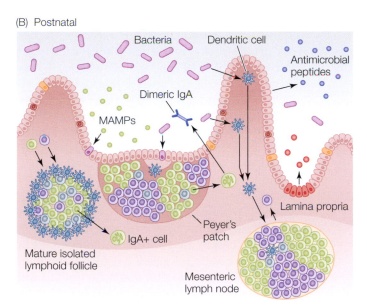

Delivery method affects disease risk

The types of bacteria that colonize the infant gut differ with delivery method. Vaginal delivery leads to initial colonization by aerobic *E. coli* and *Streptococcus* followed within two weeks by anaerobic *Bacterioides*, *Clostridium*, and *Bifidobacterium*. Bifidobacteria stimulate the maturation of the cells that secrete IgA in saliva, which appears to protect against allergy. C-sections lead to more frequent colonization by *Klebsiella*, *Enterobacter*, and *Clostridium*. Meta-analyses indicate that children born by C-section have a 20% increased risk of asthma (Thavagnanam et al. 2008) and a slightly increased risk (1%–4%) of atopy and allergy (Bager, Wohlfahrt, and Westergaard 2008).

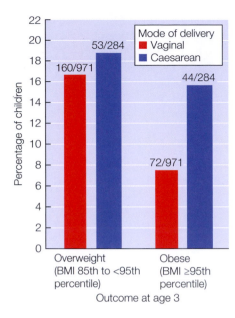

Figure 2.32   Association between mode of delivery and percent overweight and obese at age 3. *n* = 1255. (After Huh et al. 2012.)

Because gut bacteria mediate nutrient metabolism and uptake, it is not surprising that the modification of the infant microbiota by C-section also affects the risk of obesity (Huh et al. 2012). In 3-year-old children born by C-section, the risk of being obese was roughly twice that of children born by vaginal delivery (Figure 2.32).

Breast-feeding affects disease risk

Atopies are syndromes characterized by localized hypersensitivity to an allergen; they include eczema (atopic dermatitis), allergic rhinitis (hay fever), allergic conjunctivitis, and allergic asthma. Breast-feeding for at least 3 months reduced the risk of asthma from 17% to 10% ($p < 0.01$) in a birth cohort of 1218 children followed for 4 years (Tariq et al. 1998). It also slightly but not significantly decreased the risk of eczema and rhinitis.

Breast-feeding also decreases the risk of obesity. One study of 15,253 children age 9 to 14 found that exclusive breast-feeding had reduced the risk of obesity (OR = 0.66) compared to exclusive formula feeding (Mayer-Davis et al. 2006). A meta-analysis of nine studies with a total of more than 69,000 participants found a similar but somewhat smaller effect (OR = 0.78) (Arenz et al. 2004). The mechanisms mediating the effect could be either epigenetic (Figure 2.33) or a continuing difference in gut flora that could be treated with probiotics (Mischke and Plosch 2013).

Antibiotic treatments increase the risk of atopies and obesity

Any antibiotic treatment in the first three years of life doubled or more than doubled the risk of hay fever and eczema in children in the United

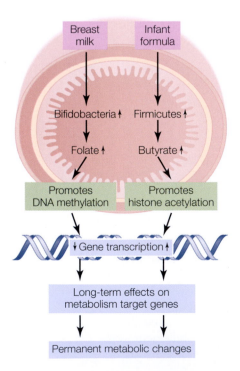

Figure 2.33   By changing the gut flora, breast-feeding may initiate a cascade of effects that lead to epigenetically regulated differences in metabolism. (After Mischke and Plosch 2013.)

Kingdom (Farooqui and Hopkin 1998). Treatments at ages 4 to 5 did not consistently increase risk. Antibiotic treatments in the first 6 months of life increased the risk of obesity slightly but significantly in children up to age 7 years (Trasande et al. 2012). Treatments later in infancy did not have consistent effects.

Antibiotic treatment can flip the gut ecosystem between alternative stable states. Prior to treatment, the gut microbiota are in a healthy stable state, species rich, with complex community metabolism and resistant to colonization by microbes that are adapted to a disturbed gut. After treatment, the gut microbiota are in a degraded stable state, species poor, with simpler community metabolism and resistant to colonization by microbes adapted to a healthy gut. In both states, positive feedbacks maintain community stability (Lozupone et al. 2012). Fecal transplants appear to be capable of reversing the degradation of the gut microbiota and restoring the healthy state.

Geographic comparisons of populations

Physicians have long known that one sees autoimmune diseases, such as type 1 diabetes, in developed countries but not in the developing world, where children grow up in environments that expose them to more parasites and pathogens. We discuss this situation in more detail under the old friends/hygiene hypothesis when we expand on mismatch in Chapter 8. Here we mention just one study to connect that issue to the development of the microbiota.

The people of Finland and Russian Karelia are genetically similar, but they live in environments that differ dramatically in hygiene and income. Finland is relatively rich (2001 average income $25,000) and cleaner; Karelia is relatively poor (2001 income $1,600) and less clean. Only 5% of Finns are infected with *Helicobacter* in contrast to 73% of Karelians, and there are similar differences in hepatitis A, *Toxoplasma*, enteroviruses, and gut nematodes (Kondrashova et al. 2012).

The incidence of type 1 diabetes is roughly two times higher (Figure 2.34) and that of Crohn's disease is nearly five times higher in Finland (1:107) than in Karelia (1:496).

That the association of hygiene with the incidence of autoimmune disease can be so strong suggests that experimental studies of mechanisms should be

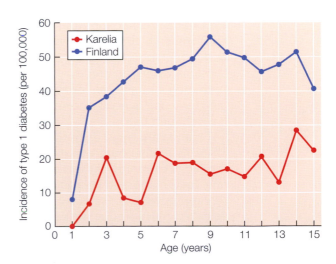

Figure 2.34 Incidence of type 1 diabetes by age in Karelia and Finland. (After Kondrashova et al. 2012.)

worthwhile, and, as we will see below, many have been done. The candidate mechanisms include competition, regulation, and tolerance. If competition is involved, pathogens are eliciting stronger immune responses that compete with the weaker responses elicited by allergens and autoantigens. If regulation is involved, the suppressive effect of one antigen also suppresses immune responses to other antigens; or, more generally, pathogens are manipulating regulatory cells to improve their own survival with the side effect of dampening the autoimmune and allergic responses of the adaptive immune system. If tolerance is involved, pathogens are steering the immune system toward tolerance rather than resistance (more on that in Chapter 4). These mechanisms are not mutually exclusive.

■ SUMMARY

The development and expression of our immune system has coevolved with our microbiota, which form the majority of cells in our body. Interventions that perturb our microbiota increase the risks of atopies, autoimmune diseases, and obesity. These interventions include C-sections rather than vaginal delivery, infant formula rather than breast milk, antibiotic treatments early in life, and hygiene. In all these cases, the result is a mismatch of organism to environment that produces diseases of civilization caused by the inability of biology to evolve as rapidly as cultures can change.

So, what *is* a patient?

Patients are mosaics of traits with very different evolutionary ages. Some of the oldest traits, such as intermediary metabolism, are very precise and efficient because they have been shaped by billions of selective events. They are also very hard to change. Some of the most recently evolved traits have not yet had time to adapt to changed conditions; they are involved in mismatches.

Patients with different recent evolutionary histories have genes that experienced selection from exposure to different diseases, diets, and environmental toxins. They vary genetically in their ability to resist disease and to metabolize drugs.

Humans have only recently evolved highly dependent young, short interbirth intervals, and menopause. These traits are not yet as precisely and elegantly designed as ancient traits because many fewer selective events have shaped them.

Patients are bundles of trade-offs. Changes in one trait are linked to changes in other traits genetically, developmentally, and physiologically. It is difficult to improve fitness or health by changing one trait without incurring costs expressed in changes in other traits.

Patients were not selected to age and die. They age because they were selected for reproductive performance, and reproductive performance early in life has benefits for fitness that outweigh the costs that it imposes in poor maintenance and increased mortality later in life.

Patients are products of their individual developmental histories as well as their evolutionary histories. They react to diet, exercise, and disease with individually variable plastic responses. Events occurring very early in life—at implantation, in utero, perinatally, and in infancy—have large and lasting effects on the rest of life. Nutrition in utero, in infancy, and in childhood affects the risk of cardiovascular disease, type 2 diabetes, and obesity decades later.

Patients have a microbiota that also develops and is affected by events occurring early in life. Choosing a C-section rather than vaginal delivery, infant formula rather than breast milk, and antibiotic therapy for infants increases the risk of asthma, eczema, autoimmune diseases, and obesity.

What Is a Disease?

The question of what is a disease can be answered several ways. Diseases consist of syndromes of symptoms, and those symptoms have causes. Answers about causes can use mechanistic reasoning, evolutionary reasoning, or both, and they can differ with the type of disease.

Classifications of disease

Diseases have traditionally been categorized as infectious, genetic, or degenerative, or by organ systems. No existing classification system adequately combines mechanisms with evolutionary insights, however, because many diseases are caused by interaction effects, primarily genotype × environment (G × E) interactions and gene × gene interactions (epistasis) that in turn also interact with the environment. The disease state of the patient is thus determined by complex contingencies. The complex relationship between genotype and disease risk in a given environment is visualized in Figure 3.1.

In Figure 3.1, every point on the curve represents some individual genotype and the associated disease risk. Some genotypes produce disease in all environments, and some environments (here environment D) can cause some risk of disease for almost any genotype. This graph represents a specific disease and a small set of environments; analogous images with different risks and thresholds would be needed to represent other cases. The same type of representation can be made for individual genetic variants—alleles at a single locus—instead of genotypes.

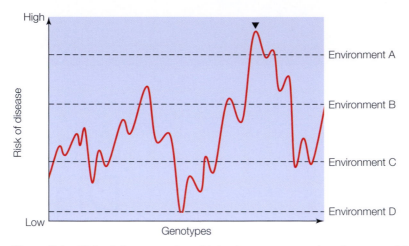

Figure 3.1 Risk of disease varies with both genotype and environment. Like many other phenotypic traits, disease risk is a complex G × E interaction. The black triangle calls attention to a genotype that produces disease in every environment.

To provide a structure that reflects both evolutionary causes and proximate mechanisms and to set up the discussion of defenses in Chapter 4, we offer the classification system given in Table 3.1. Rather than using Aristotelian logic that makes clean distinctions among sharply defined categories based on discrete attributes, this system is based both on fuzzy logic that recognizes that disease states are risks rather than deterministic consequences of prior conditions and on continuous attributes, such as the degree of genetic or environmental impact. Some diseases belong to more than one category. A disease appearing in more than one category—for example, type 2 diabetes is both a disease of homeostasis and a disease of maintenance—signals that the disease can have more than one cause. It is the causes, both evolutionary and mechanistic, that are emphasized in these categories.

TABLE 3.1 A classification of diseases that explicitly recognizes complex causation

Category	Diseases
1	Diseases with genetic causes
2	Diseases with environmental causes
3	Diseases that are by-products of defense systems
4	Diseases of homeostasis
5	Diseases caused by lack of maintenance
6	Diseases caused by stochastic developmental problems
7	Diseases of pregnancy and early development caused by maternal-fetal and maternal-paternal conflicts

Genetic diseases (category 1) are usually caused by de novo or recessive mutations to single genes. They can affect any aspect of biology, from metabolism to behavior, and are rare and catastrophic, although not severe enough to be lethal during early embryonic development. Their deleterious effects can cause disease in any environment and are normally lethal or prevent reproduction. Usually, no specific mechanisms have evolved to protect us from them. The prevalence of diseases in this category slightly increases with medical progress because more of the cases that had previously been lethal can now be rescued. Gene therapy is currently considered the only therapeutic option for most diseases in this category, which include cystic fibrosis, severe combined immunodeficiency, and thalassemias. Phenylketonuria, however, can be prevented by a special diet devoid of phenylalanin.

In direct contrast, because *environmental diseases* (category 2), as major contributors to extrinsic mortality, are among the leading causes of natural selection, we have evolved mechanisms to deal with them. The more prevalent they were in our evolutionary history, the stronger the genetic component of the mechanisms used to deal with them. Among the diseases in this category, there is a gradient from genotype-independent, catastrophic environmental insults to milder and more common environmental stressors whose induction of disease can depend on genotype. Environmental diseases include infections; dysbiosis (disruption of symbiosis with our normal microbiota); starvation and malnutrition (caloric, vitamin, and micronutrient deficits); dehydration and electrolyte imbalance; hypoxia; hyper- and hypothermia; mechanical injuries (predation, conflict, accidents); and exposure to venoms, toxins, poisons, pollutants, and other noxious xenobiotics. Modern environments reduce the prevalence of some of these diseases (e.g., infections and starvation), but in so doing, they increase the prevalence of category 3 diseases.

Diseases that are by-products of defense systems (category 3) result from the costs of defense, from deficient and excessive defenses, and from malfunctions of defense systems. Usually, these vulnerabilities in our defense systems are exposed by environmental mismatches, and, like the diseases of maintenance (category 5), many are ultimately caused by a conflict between survival and reproduction on the one hand and health and well-being on the other. The prevalence of these diseases has grown as the prevalence of category 2 diseases has declined. They include immunopathologies, autoimmune diseases; allergic diseases; asthma; fibrosis and metaplasia; hemodynamic and thromboembolic diseases; and obsessive-compulsive disorder, phobias, paranoia, anxiety, and panic disorders.

Diseases of homeostasis (category 4) are also exposed by environmental mismatches. They affect physiological systems that were designed to be plastic and adjustable, which makes them vulnerable to disregulation. They can result in part from inappropriate adjustment of homeostatic set points. Phenotypic plasticity is a developmental counterpart of adjustable homeostatic set points. Both homeostasis and plasticity afford flexibility in the face of varying environments or physiological priorities, and both enable diseases when the system is set and locked in the wrong state. Homeostasis can be maintained by both physiological and behavioral mechanisms, which share a liability. Addictive behavior can be viewed as a consequence of set point change in the pleasure centers of the nucleus accumbens, just as obesity can be viewed as a change in the hypothalamic set point for control of body weight. The diseases of homeostasis include obesity, type 2 diabetes, atherosclerosis, hypertension, other cardiovascular diseases, and addictive behavior.

Diseases of maintenance (category 5) are caused not by the presence of an insult but by the absence of maintenance. Unlike inducible defenses, which are triggered transiently by specific threats, maintenance processes operate continuously to deal with by-products of normal physiology and to sustain normal operations. Decline of maintenance over time results in diseases of aging, which affect structures and processes that need the most maintenance and that are most costly to maintain. These diseases are mediated by antagonistic pleiotropy, if we are thinking of genetic causes, or by trade-offs, if we are thinking of physiological causes. Both types of causes contribute. In a state of evolutionary equilibrium, the intrinsic mortality caused by the decline of maintenance is in rough balance with extrinsic mortality; therefore, these diseases increase in frequency when a relatively sudden drop in extrinsic mortality extends life span. Maintenance programs can prevent diseases in this category, but no mechanisms have evolved to defend against them once they occur. For example, autophagy can effectively prevent some neurodegenerative diseases, but once these diseases occur, there is no evolved mechanism in place to cure them (in the sense that the immune system can cure many infectious diseases). The diseases of maintenance include degenerative diseases (neurodegeneration and sarcopenia), age-related cancers (the majority of cancers), osteoporosis, atherosclerosis, cardiovascular diseases, type 2 diabetes (those last three can also be caused by loss of homeostasis), and arthritis.

Diseases caused by stochastic events resulting in developmental problems (category 6) are consequences of trade-offs and random events whose causal basis is not yet well understood. They are either too rare or too expensive to fix. They include random chromosomal segregation errors resulting in trisomy 18 and 21, fetal hydrops, diseases of premature delivery (neonatal respiratory distress syndrome, necrotizing colitis, jaundice), and sudden infant death syndrome. Most of the diseases in category 6 are prevented by early spontaneous abortions and oocytic attrition. Some of them, such as pre-eclampsia, have complex causes that also confer membership in category 7.

Diseases caused by disruptions of equilibria in maternal-fetal and maternal-paternal conflicts (category 7) include diseases of pregnancy, such as pre-eclampsia and gestational diabetes, and early development as well as some components of the causation of mental disorders like autism and schizophrenia. Fetal growth restriction and blood group incompatibility also belong in this category.

Four perspectives on human diseases

Although the distinction between health and disease seems simple and obvious, it looks different from four perspectives: those of a patient, a physician, an evolving pathogen, and an evolving host. From a patient's perspective, disease is a state of "not feeling well": pain, nausea, fatigue, discomfort, and depression are symptoms we associate with sickness. The patient is more concerned with symptoms than causes. From a physician's perspective, the nature of a disease is much more nuanced and dependent on current medical knowledge. In general, physicians view diseases as abnormal conditions that result in symptoms that compromise the health of patients and want to know immediately what caused them.

The evolutionary perspective can take the view of the pathogen or the host. From a pathogen's evolutionary perspective, the symptoms of its host's disease can be an adaptation if they promote its reproduction or transmission; or, more commonly, they can be an unintended by-product of the pathogen's interactions with the host. From the host's evolutionary perspective, diseases can be both by-products of natural selection (degenerative and chronic diseases mediated by antagonistic pleiotropy) and substrates for natural selection (when they affect reproductive success), and many symptoms of disease are defense mechanisms in action.

For example, from the patient's perspective, diarrhea causes pain, discomfort, and inconvenience; from the physician's perspective, it can cause severe dehydration with secondary consequences; from the pathogen's evolutionary perspective, it can be a means of promoting transmission; and from the host's evolutionary perspective, it can be a defense that expels pathogens causing intestinal infections. The distinction between health and disease can thus depend on one's perspective.

Diseases have both mechanistic and evolutionary explanations

All biological phenomena have both mechanistic and evolutionary explanations. Evolutionary and mechanistic explanations are complementary, not exclusive, because they are given at different levels. Both contain clues for treatments.

For example, Crohn's disease is a type of inflammatory bowel disease that presents with abdominal pain, diarrhea, and weight loss and can be caused by deregulation of the immune response to intestinal microbiota. Mechanistic explanations of Crohn's disease are couched in terms of genetics, development, and the biochemistry of nucleotide-binding oligomerization domain (NOD) receptors, regulatory T cells, and anti-inflammatory cytokine IL-10. In particular, the NOD2 protein, which is located in the cytoplasm of the intestinal epithelial cells and macrophages, senses the presence of bacterial peptidoglycans and regulates the innate immune response to infections. GWAS (genome-wide association studies, see Chapter 2) of NOD mutants report that homozygous mutants have a 17-fold greater risk of Crohn's disease and that heterozygous mutants have a 2.4-fold greater risk (Cho 2008). This explanation combines genetics with biochemistry.

Evolutionary explanations of Crohn's disease are based on our previous coevolution with a community of microbiota that were stable for a long time; we responded to them with many developmental and physiological changes. When that community was suddenly, recently, and drastically altered, our development and physiology responded maladaptively to a new environment for which they did not have any evolutionary experience and to which they have not yet had time to adapt. The severity of the resulting symptoms appears to depend on the patient's microbiota.

Reconstituting some of the signals that were reliably present in the gut environment over evolutionary history but that are missing today is one potential approach to therapy. We all used to be infected with worms that manipulated our immune systems, and our immune system adapted to them. Removing worms relieves the symptoms of worm infections, but it also produces an abnormal immune response. It is an example of a mismatch between slow-moving biology—our evolutionary history of worm infections—with hygiene produced by rapid cultural evolution. Not all parasitic worms are desirable candidates for therapy because most of them produce lasting infections with serious symptoms. Worms that are related to human parasites but that have not evolved on humans may be better candidates because they are not well enough adapted to the human environment to produce a persistent infection, but they may produce signals similar enough to mimic the effects of more serious parasites. One such worm is the pig whipworm, *Trichuris suis*, a nematode a few centimeters long and about the diameter of a thick hair. Therapy with pig whipworm eggs has looked promising in preliminary trials.

Genetic and environmental causation are each of two types

The two types of genetic causation are defects and predispositions. Genetic defects are relatively rare catastrophes caused by single mutations (category 1 in Table 3.1), such as Duchenne muscular dystrophy or severe combined immunodeficiency. Most human genetic diseases are rare defects caused by mutations that are unconditionally detrimental. Important exceptions are genetic diseases caused by allele variants that confer advantage in specific environments. One example of this is sickle cell disease, where one allelic variant of hemoglobin has been driven to high frequency in populations residing in areas where malaria is endemic.

Genetic predispositions are often polygenic and involved in environmental interactions. They are often neutral or beneficial in some settings but detrimental in others. When the environments in which they are beneficial occur frequently, the genetic variants increase in frequency in response to selection. For example, a particular genetic inability to process ethanol caused by variation in alcohol dehydrogenase leads to the rapid buildup of aldehydes, which causes unbearable hangovers, but it may protect against toxins produced by fungi that infect stored rice (Han et al. 2007). Here, the type of outcome associated with a genetic variant depends on what the patient is ingesting.

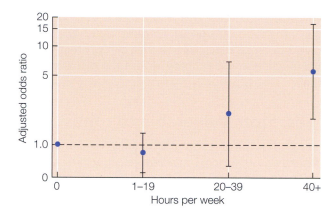

Figure 3.2 The risk of peripheral arterial disease increases with exposure to secondhand smoke. Confidence bounds represent the 95% confidence interval. Note nonlinear axes. (After Lu, Mackay, and Pell 2013.)

Environmental causation has a similar structure. Environmental catastrophes can lead to serious injury or death and include accidents, attacks by predators, and infection with highly virulent pathogens such as Ebola or *Vibrio cholerae*. Accumulative environmental effects occur with smoking, overeating, and lack of exercise, and they affect the risk of cardiovascular disease, cancer, and type 2 diabetes. For example, the risk of peripheral arterial disease increases significantly with the number of hours per week that a person is exposed to secondhand smoke (Figure 3.2).

Gene-environment interactions: G × E causation

The second types of genetic and environmental causation both involve G × E interactions. When an effect is produced by an interaction of causes, it is a mistake to assign responsibility to just one component because the processes that cause the effect to appear are interactions of components (Hilborn and Stearns 1982).

For example, the environmental agents that affect the risk of cancer, such as smoking and air pollution, have effects mediated by genetic variation in the *N*-acetyl transferases (NATs). Patients with the combination of an NAT2 slow acetylator with an NAT1 fast acetylator and who smoked were 2.73 times more like to get bladder cancer (Cascorbi, Roots, and Bröckmoller 2001). Patients with both NAT1 and NAT2 fast acetylator alleles and who smoked were at increased risk of colorectal cancer (Sørensen et al. 2008). Note that in these cases the interactions are both gene × gene (epistatic) and gene × environment: here it takes a combination of particular alleles at two loci in interaction with an environmental factor to increase risk of cancer.

■ SUMMARY

Diseases have both mechanical and evolutionary explanations. Genetic causes can act either catastrophically or through predispositions. Environmental causes can act either catastrophically or through accumulative effects. Many diseases result from G × G and G × E interactions that are missed by simple categories. Important G × E interactions often occur when genetic effects are polygenic and environmental effects are accumulative. Based on evolutionary and mechanistic causation, we place human diseases into seven categories. Depending on the causation, organisms evolved defense mechanisms that can protect from a disease, maintenance mechanisms that help prevent the disease, or no such mechanisms at all.

Organs and tissues: From vulnerable to robust

Organs and tissues vary in their degrees of robustness, resilience, and vulnerability. Let's consider a tissue, organ, or the entire organism as being in either the healthy state or in one of many possible unhealthy states. Robustness is a characteristic that defines how resistant that system is to transition from the healthy to an unhealthy state, and resilience is a characteristic that defines how easily the system can return from an unhealthy state back to the healthy state. Robustness and resilience do not necessarily go hand in hand and, in fact, are often negatively correlated. Bones, for example, are robust in terms of resistance to mechanical damage but have low resilience in terms of recovery from damage. Conversely, intestinal epithelium and skin have low resistance to damage but are very resilient, and damage to these tissues can be repaired relatively quickly and easily. What is hard to damage is hard to repair, and what is easy to damage is often easier to repair.

The problem is with the tissues and organs that are easy to damage and hard to repair, such as the brain, lungs, kidney, and heart. These "weakest links" in the body define the most important vulnerabilities to morbidity and mortality. The mosaic of organ and tissue robustness and resilience determines the spectrum of diseases, both in general and in response to particular environments. Not all organs and tissues are as robust and resilient as possible because of trade-offs: producing the necessary structures and processes would cost more than it would pay. In the following sections, we explore the issues of renewal and repair, sensitivity and stress, vulnerability to threats, what determines the weakest physiological links, built-in safety factors, trade-offs and constraints, and trade-offs and mismatch.

Tissue renewal and repair

Tissues with a high renewal rate and repair capacity, such as most epithelia and the hematopoietic system, are relatively tolerant to damage. Tissues with low renewal rate and repair capacity, such as neurons and cardiomyocytes, do not tolerate damage. Damage to the former is easily handled, whereas comparable damage to the latter can be fatal. That does not complete the list of costs and benefits, however, because tissues with high renewal rate and repair capacity are much more susceptible to developing cancer—another trade-off and a vulnerability.

Sensitivity to stress

Tissues with high energetic demands, such as neurons, heart muscle, and kidney tissue, rely exclusively on oxidative metabolism. Tissues with lower energetic demands, such as adipocytes, fibroblasts, and hematopoetic cells, tolerate hypoxia relatively well because they can switch from oxidative to glycolytic metabolism. That is why ischemia—a lack of oxygen and nutrients—rapidly and seriously damages brains, hearts, and kidneys, often irreversibly and sometimes fatally.

Vulnerability to threats

Protective structures have evolved to shield vulnerable organs from frequently encountered threats: the skull protects the brain; the rib cage protects the heart and lungs. Some organs, including the brain, the eyes, and the gonads, are protected from inflammatory damage by suppression of local immune responses. Those organs pay a price for avoiding the costs of inflammation and are vulnerable to the pathogens that manage to gain access to them, a trade-off.

The weakest physiological links

The consequences of loss of function vary widely among tissues and organs. Mild damage to liver, skin, or intestinal epithelium can be tolerated and repaired; it is often not lethal. Equivalent damage to the respiratory, cardiovascular, or central nervous system causes severe disease and even death. These systems are the weakest links in the body. Their malfunction kills; their failure is a frequent proximal cause of death, whether from infection, trauma, or old age. It is therefore in the weakest links where special features have evolved to protect against catastrophic failure. One way to conceptualize them is as safety factors.

Built-in safety factors

The safety factor is the ratio of functional capacity to expected maximum load. If body weight is W and the skeleton can bear a weight of $4W$, the safety factor of the skeleton is 4. Increasing the safety factor has costs. In this case, further strengthening the skeleton would decrease mobility. In general, the greater the safety factor, the greater the resistance to damage but the larger the costs of that resistance.

Here are some estimated safety factors for biological materials (Weibel, Taylor, and Bolis 1998):

- Human pancreas: 10
- Wing bones of a flying goose: 6
- Human kidneys: 4
- Leg bones of a running ostrich: 2.5
- Human small intestine: 2
- Backbone of a weightlifter: 1.0–1.7

Trade-offs and constraints explain vulnerability

Many structures are vulnerable because of compromises imposed by trade-offs. The synovial joints—wrist, elbow, finger, shoulder, and knee—have superior mobility but are vulnerable to arthritis. The alveolar sacs enable excellent gas exchange but are vulnerable to pneumonia when filled with inflammatory exudate.

Other structures are vulnerable because of historical constraints. The vertebrate eye has a blind spot where the axons, which in vertebrates lie on the distal side of the retina, coalesce and travel into the brain. The octopus does not have that blind spot because its eyes develop with the axons proximal to the brain. Mammalian sperm need to develop at temperatures lower than core body temperature; the testes therefore descend into an external scrotum, where they are at greater risk of damage than they would be if carried internally.

Trade-offs and mismatch: When large benefits can carry large costs

After a sufficient amount of evolution in a stable environment, the cost-benefit balance of trade-offs will be optimized for traits important to fitness. Traits that are capable of making large contributions to fitness can then be involved in trade-offs that contain large potential costs. In such cases, the effects of disturbances can be serious and pathologies can result, either because the mechanisms controlling the cost-benefit balance of the trade-off are disturbed or because the environment changes. For example, the immune defense against infection, a large benefit, brings with it the large costs of immunopathology and the risks of sepsis, anaphylactic shock, and autoimmune diseases. The clotting system, a large benefit when wounded and bleeding, carries with it the risk of embolism and stroke.

Environmental change can alter the balance of costs and benefits. Sometimes, the costs exceed the benefits until evolution catches up, and that takes time. Modern hygienic environments are abnormal; they elicit pathological immune responses that include asthma, allergies, eczema, and autoimmune diseases. Similarly, cultural changes in diet combine with inflammation to increase the risk of clotting disorders.

■ SUMMARY

The morphology and physiology of the patient's body is a mosaic of tissues, organs, processes, and structures that vary in vulnerability to damage, capacity for repair, sensitivity to stress, and built-in safety factors. Each of these features has benefits and costs; each is involved in trade-offs. When perturbations shift the cost-benefit balance, they elicit pathologies. In this sense, pathologies are situations in which costs exceed benefits, just as normal function is defined by situations in which benefits exceed costs. Trade-offs thus define a balance that can be tipped from normal function to pathology.

From fixed to adjustable

Some physiological functions, such as glucose metabolism and blood pressure, are very sensitive to environmental change. Others, such as DNA replication, are insensitive. The sensitive functions have evolved to be flexible:

TABLE 3.2 Contrasting syndromes of fixed and adjustable functions

Fixed functions	Adjustable functions
Insensitive to the environment	Sensitive to the environment
Highly conserved across the tree of life	Vary within and between species; constantly shaped by natural selection; directly involved in local adaptation
Control core biological processes	Control physiological adaptation, inducible responses, reaction norms
Disrupted by single catastrophic mutations, responsible for rare Mendelian diseases (e.g., cystic fibrosis, progeria, severe combined immune deficiency)	Affected by many genes, often polymorphic, whose frequencies change under selection; cause predisposition to diseases of mismatch (e.g., obesity, type 2 diabetes, asthma, hypertension, atherosclerosis)

they can adjust to changing environments and physiological priorities. The functions that are insensitive to the environment tend to be developmentally and functionally invariable, with fixed characteristics. The same distinction broadly applies to phenotypic traits: some are fixed (e.g., body plan); others are adjustable (e.g., body weight). The adjustable functions buy their flexibility at the cost of vulnerability to disregulation. For example, the flexibility of regulation of glucose metabolism created our vulnerability to insulin resistance and type 2 diabetes. In contrast, insensitive functions are robust to such disregulation but remain vulnerable to genetic defects. Each category of function—fixed versus adjustable—forms a syndrome of characteristics (Table 3.2).

Fixed functions include core biological processes like DNA replication, transcription, cytoskeleton function, and major features of the body plan, such as the number of heads, legs, arms, digits, and eyes. Mutations disrupting such traits can have catastrophic, lethal effects. They are often eliminated during development by quality control of gametes, zygotes, and concepti. When not eliminated early in development by natural quality-control devices, they result in major birth defects, such as anencephaly, ectrodactyly, and Patau syndrome.

One example of an adjustable process is glucose metabolism. Glucose allocation among different tissues (e.g., skeletal muscle, adipose, liver) is regulated by insulin, which is produced by pancreatic β-cells in response to elevated blood glucose level. As in all homeostatic systems, the level of glucose is maintained by negative feedbacks that keep the glucose level close to the set point. Under some conditions, however, the set point for glucose homeostasis needs to change to accommodate different priorities or to adjust to different environments. The change in set point occurs in part through alteration in the insulin sensitivity of target tissues. For example, during pregnancy, maternal tissues can become insulin resistant to allocate more glucose to the fetus, which has higher physiological priority. Likewise, during an infection, glucose is reallocated from muscle and fat to the immune system by making muscle, fat, and liver insulin resistant. These changes in set points are beneficial, but the ability of the system to change the set points creates the vulnerability to disregulation and diseases of homeostasis, in this case, type 2 diabetes.

■ SUMMARY

Fixed traits are insensitive to the environment, deeply conserved, and often developmentally canalized. Mutations disrupting such traits often act early in development with catastrophic consequences. Adjustable traits, which react sensitively to changes in the environment, are often controlled by many genes with allelic variation. The disruption of adjustable traits, whether by genetic mutations or environmental insults, produces diseases of homeostasis.

The changing nature of disease

Disease prevalence (the fraction of a population afflicted with a given disease) and disease incidence (the number of newly diagnosed cases per time period) can vary widely among regions of the world. The prevalence and incidence of some diseases can also change dramatically over relatively short periods. Some diseases tend to occur in clusters in which high prevalence of one disease is associated (either temporally or geographically) with high prevalence of another disease. In some cases, it is for trivial reasons; for example, high prevalence of two co-occurring diseases could result from poor nutrition or lack of medical care. In other cases, the correlations (positive or negative) in prevalence or incidence of diseases have biological reasons reflecting genetic and physiological trade-offs that expose human vulnerabilities. The disease classification introduced at the beginning of this chapter and summarized in Table 3.1 provides a natural framework for exploring the changing patterns of human disease.

Monogenic and stochastic developmental diseases

For the most part, the incidence of monogenic diseases and stochastic developmental diseases is independent of the environment except when the environmental factor has a mutagenic activity (e.g., UV light exposure, radioactivity, chemical mutagens) or can increase the rate of stochastic developmental problems (e.g., during pregnancy). Variable presence of mutagens in the environment (or variable exposure to environmental mutagens such as UV light) can and does change the prevalence of some diseases in time and space. In general, however, the rare mutations that cause monogenic diseases occur at random and at a constant rate. Although the incidence of these diseases may remain relatively unchanged, their prevalence can increase due to medical interventions that prolong or enable survival of genetic disorders that were lethal just a few decades ago. Thus, the rate of occurrence of mutations causing phenylketonuria has likely been the same for millennia, but survival of patients with this disorder became possible only recently. Accordingly, the prevalence of these and other diseases is growing with the advent of medical interventions that permit survival of the patients afflicted with more and more monogenic diseases. When medical interventions increase survival sufficiently and the mutations that cause the disease are compatible with reproduction, we can expect that the frequency of the mutant alleles will increase over the levels found in the past, when survival or reproduction with the same genetic defects was not possible.

Environmental diseases

Diseases considered environmental diseases by definition change with the environment. Major environmental transitions in human evolutionary history include exit from Africa, the agricultural revolution, the industrial revolution, and the associated transitions: high-density settlements, urbanization, long-range travel, introduction of sanitation and hygiene products, clean water, advent of electricity and change of the light cycle, processed food, infant formula feeding, and the advent of pharmaceuticals and broadly used medical interventions (antibiotics, vaccines, nonsteroidal anti-inflammatory drugs, statins, painkillers, and histamine blockers). These transitions have dramatically affected the incidence of many human diseases, ranging from infections and allergies to obesity and behavioral disorders. Diseases caused by environmental changes are often a consequence of a mismatch between evolved human biology and modern environments, a subject discussed in greater detail in Chapter 8.

Here, we emphasize that the outcome of a mismatch depends on whether or not the disease in question involves a trade-off. Some new diseases emerge as a direct consequence of novel environmental toxins or pathogens, such as dioxin poisoning and AIDS. If they persist, they can drive natural selection in human populations; in theory, however, they can be completely eliminated by removing the toxin from the environment or by developing a vaccine against the pathogen. Other diseases become more prevalent when an environmental change shifts the cost-benefit balance of a trade-off that was optimized for previous environments. Examples include diseases of the defense systems. Immune responses selected in past epidemics, for example, may increase the risk of diseases caused by the costs of immunity, including autoimmune diseases, allergies, and asthma (see Figure 8.3). Changing the environment to eliminate the latter set of diseases is not usually an option because doing so would reintroduce the very conditions—such as infections or malnutrition—that selected the gene variants that now confer disease risk.

Changes in the cost-benefit balance of trade-offs are also now responsible for the high prevalence of many diseases of homeostasis. Type 2 diabetes and cardiovascular diseases are among them.

The transition to modern environments has caused a dramatic decline in the extrinsic mortality caused by childhood infections, starvation, and other hostile environmental factors. The resulting extension of life span has exposed many age-associated maladies that were rare until recently but are now common sources of intrinsic mortality. These diseases include cancer, neurodegeneration, and diseases of failed defenses, all caused by a decline in maintenance programs. If these diseases are conquered by medicine, new ones will emerge that will become the common causes of intrinsic mortality. How much the average human life will be extended is unclear.

■ SUMMARY

The spectrum of common human diseases has been changing dramatically. The incidence of diseases caused by random processes (somatic mutations or birth defects) is relatively unaffected by the environment, but their prevalence can increase when reduced extrinsic mortality caused by medical interventions increases life span. Most changes in human disease patterns, however, result from the rapidly changing environment. Some of them are caused by the introduction of new environmental factors (industrial toxins, trans fats, emerging pathogens), and others are due to disruptions of trade-offs in which past selection processes increased the current risk of diseases of defenses, homeostasis, and reduced maintenance.

Defenses

Defenses constitute a set of diverse processes among which distinctions must be drawn if we are to understand the consequences of perturbing them. Both the evolutionary origins and the current state of defense mechanisms determine the functions with which they trade off. Two major themes run through this chapter. First, defense mechanisms have usually evolved through modification of processes that evolved earlier as part of normal physiology. Second, evolution will tolerate high costs and permit significant vulnerabilities when those costs and vulnerabilities are associated with major benefits. Because some defense mechanisms yield major immediate benefits, they can also be associated with high costs and significant vulnerabilities.

Maintenance and defense originated in homeostasis

Homeostasis is the stability of key regulated variables in cells, tissues, and organisms. It is made possible by mechanisms that evolved to detect variation in those key variables and activate feedback processes to restore them to appropriate levels. Most maintenance and defense mechanisms are evolutionarily derived through modification of processes that were originally homeostatic. *Maintenance mechanisms* sustain the normal functioning of cells and tissues and prevent, repair, and reduce the malfunctions that result from deteriorations that are the by-products of normal biological processes. *Defense mechanisms* protect from specific hostile environmental factors—perturbations that exceed the homeostatic capacity of the organism (Figure 4.1).

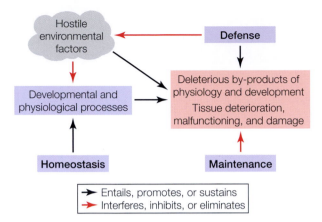

Figure 4.1 Homeostasis, maintenance, and defense are interrelated.

Distinctions among the three processes

To illustrate the distinctions among homeostasis, maintenance, and defense, we consider two examples.

First, the protein concentration in a cell is regulated by homeostatic mechanisms through control of protein synthesis and degradation. Damaged (oxidized, aggregated, or misfolded) proteins are continuously produced even under perfect homeostatic conditions. They are by-products of normal cellular physiology that are detected by specialized sensors, which promote their clearance and degradation. This maintenance mechanism prevents cellular deterioration due to proteotoxicity. Viruses in cells are not a homeostatic variable and are not a by-product of normal cellular physiology. They are detected by specialized sensors of the innate immune system that induce antiviral defense. Defenses operate at the expense of homeostasis: during a viral infection, cellular protein synthesis is shut down. Defenses also increase the generation of the by-products of cellular physiology (e.g., damaged proteins), both as a consequence of suppressed homeostasis and due to unavoidable collateral damage inflicted by defenses. These by-products are again handled by the maintenance mechanisms. These maintenance mechanisms contribute to the *tolerance* component of defense, whereas the mechanisms that eliminate the virus represent the *resistance* component of defense. We say more about resistance and tolerance later in this chapter.

Second, cell number per tissue compartment is a regulated variable that is maintained by homeostatic mechanisms through control of cell proliferation and cell death to maintain cell quantity. When cells wear out as a by-product of their normal function, they must be continuously replaced through tissue renewal; this process is a maintenance mechanism concerned with cell quality. It has to operate continuously to prevent deterioration of cells and tissues. Tissue repair, on the other hand, is a defense mechanism that is induced on demand in response to tissue damage. It is

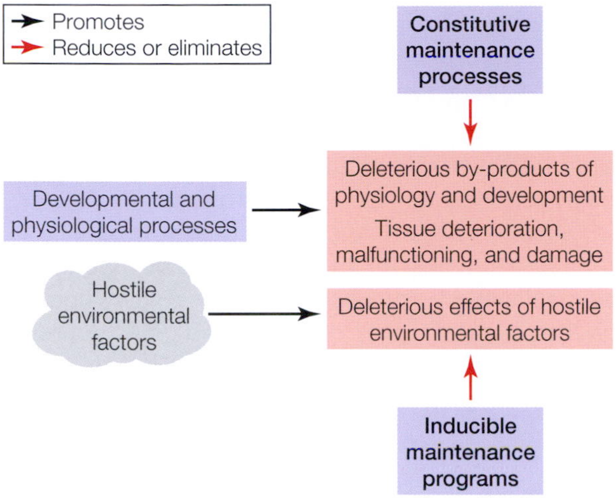
Promotes
Reduces or eliminates
Constitutive maintenance processes
Developmental and physiological processes
Deleterious by-products of physiology and development
Tissue deterioration, malfunctioning, and damage
Hostile environmental factors
Deleterious effects of hostile environmental factors
Inducible maintenance programs

Figure 4.2 Two types of maintenance programs: constitutive and inducible.

concerned with tissue integrity in the face of a damaging agent. Note that a damaging agent (pathogen or toxin) is neither a regulated variable nor a by-product of normal biological function. Tissue repair, however, a defense, is only possible in tissues that have tissue renewal through maintenance, which in turn can only operate together with homeostatic mechanisms that maintain normal cell number in tissue compartments.

Two types of maintenance mechanisms

Although the concepts of homeostasis and defense are well defined at the mechanistic level, the notion of maintenance has primarily been explored from evolutionary and ecological perspectives. Here, we expand the concept of maintenance to highlight its two essential aspects: one reflecting the role of maintenance in supporting normal developmental and physiological processes, and the other reflecting the role of maintenance in survival of harsh environments (Figure 4.2). These two functions of maintenance mechanisms complement homeostasis and defense, respectively. Some examples of the two functions of maintenance mechanisms are given below. Note that most of these constitutive maintenance mechanisms decline with age, producing various age-associated diseases.

The following are examples of constitutive maintenance mechanisms that support normal developmental and physiological functions:

- Removal of damaged proteins and lipids (proteosomal and lysosomal degradation pathways that include autophagy). Age-related decline leads to neurodegeneration.
- Detoxification and disposal of metabolic waste products. Age-related decline leads to hepatic and renal insufficiency and diseases.
- Clearance of senescent and apoptotic cells through phagocytosis. Accumulation of senescent cells in old animals can contribute to some aging phenotypes.

- Clearance of the extracellular damaged and modified proteins and extracellular matrix degradation products through phagocytosis. Age-associated decline may lead to tissue deterioration.
- Tissue renewal and regeneration through maintenance of stem cell pools. Age-associated decline can lead to degenerative diseases.
- Maintenance of tissues required for homeostasis of body temperature, blood glucose, calcium, and phosphate through brown and beige fats, skeletal muscle, and thyroid gland functions. Age-related decline results in reduced homeostatic capacity and diseases of homeostasis, such as type 2 diabetes and osteoporosis.
- Maintenance of tissues required for defenses, primarily bone marrow and thymus, where age-related decline increases susceptibility to infections.

The following are examples of inducible maintenance mechanisms that promote survival in harsh environments:

- Organismal adaptations to starvation, cold, dehydration, hypoxia, and other forms of physiological reprogramming as well as tissue protection. These maintenance programs decline with age, thus accounting for reduced adaptability to harsh environments in the elderly.
- In some organisms, suspended animation, including hibernation, estivation, dauer larvae, and sporulation. These states can involve a decline in basal metabolic rate; decreased reliance on oxygen, nutrient, and water availability; increased protection of tissues and organs; increased resistance to stress; increased tolerance to damage; and changes in the homeostatic set points to promote survival in highly hostile environments.

Inducible versus constitutive processes

Although homeostasis operates continuously, it can have inducible components. Control of cutaneous blood flow is a constitutive component of body temperature homeostasis, whereas shivering and sweating are the inducible components. Maintenance operates constitutively, but it can be up-regulated to a higher level when needed. That happens when the generation of by-products of normal physiological processes increases, for example, during exercise, exposure to infections and toxins, or as a consequence of collateral damage caused by defenses.

Constitutive maintenance processes have to operate all the time. They do not trade off with growth and reproduction in the short term within individual organisms. Inducible maintenance mechanisms operate under special physiological conditions and in hostile environments where they do detract from growth and reproduction while promoting survival. A reduction in maintenance compromises both homeostasis and defenses (hence the diverse types of diseases of maintenance in category 5; see Table 3.1).

Although in the short-term maintenance does not trade off with growth and reproduction at the level of physiology within individual organisms, in the longer term, at the level of populations, maintenance will trade off with growth and reproduction if it has a negative genetic correlation with those processes. When it does, selection to increase either reproduction or growth will have to be bought at the cost of decreased maintenance.

Defenses that are only needed on demand and have high costs are inducible, but those that are always needed—for example, barrier defenses—are constitutive. Problems normally handled by maintenance (e.g., elimination of damaged cells and proteins) are exaggerated when defenses are deployed; in that sense, maintenance is an important component of defense.

■ SUMMARY

Homeostasis, maintenance, and defense are intimately related. Maintenance and defense are both extensions of normal homeostatic mechanisms and are, in many cases, evolutionarily derived from them. Homeostasis, maintenance, and defense have constitutive and inducible components. Whether the constitutive or the inducible mode of operation is used depends both on demand and on the cost of operation.

Types of defense and costs

Defense mechanisms evolved to protect the organism from specific environmental challenges. Because the type of defense that evolves depends on the nature of the environmental hostility against which it defends, here we review the types of environmental hostility. In constructing any defense, the major design consideration is that all defenses have costs: they participate in trade-offs. Defenses may therefore allow disease for two general reasons: either they fail to protect or the costs they incur are excessive. Because evolution acts to increase the frequency of genetic variants, often at the expense of individual health and survival, costs that are evolutionarily acceptable to genes may be culturally unacceptable to patients and physicians. Some costs of defense can be the main cause of disease symptoms.

The nature of environmental hostility

The environment consists of many factors that affect reproductive success; some do so directly, others indirectly. Environmental factors like oxygen, food, water, temperature, and pathogens are universal; factors like predators and toxins are specific to particular organisms. Despite the long evolutionary history shared by all extant organisms—more than 3.5 billion years—natural environments remain challenging, with some environmental factors usually being suboptimal and at times capable of threatening life. They are the hostile factors. They may be present continuously, occasionally, or periodically (daily or seasonally). When hostile factors have sufficient impact on reproductive success, they cause the evolution of specialized defense mechanisms. Like the hostile factors against which they evolve,

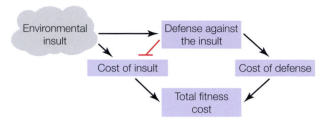

Figure 4.3 Total fitness cost is the sum of the cost of the insult (e.g., pathogen-induced damage) plus the cost of the defense.

some defense mechanisms are universal, whereas others are species- or clade-specific.

Defenses have costs: They participate in trade-offs

Defense mechanisms evolved to protect organisms from challenges such as starvation, infection, and predators. Such challenges select maintenance programs that promote survival at the expense of growth and reproduction, up to a point. Selection operates on variation in reproductive success, not just on survival, and improvements in maintenance will cease when they no longer contribute to increases in reproductive success. At that point, maintenance will be imperfect, the body will have some residual vulnerability to the environmental challenge, and life span will be limited.

Specialized defenses trade off with growth and reproduction as well as other physiological functions. Because defenses are essential for reproductive success, they have large benefits. That is important because it means that when the level of defense is optimized by selection operating in a framework of trade-offs, the acceptable costs can also be very large. Thus, *large benefits permit large costs.*

Defense mechanisms are designed to minimize the negative impact of environmental challenges. The greater the environmental insult, the greater the expression of the corresponding defense and with it the cost of that defense. Thus, the total fitness cost of an environmental challenge includes both the cost of the insult and the cost of the defense against that insult (Figure 4.3).

This point has two implications. First, defenses are constrained by their costs: the optimal defense is not the maximal defense. Second, because it is the total cost that matters, minimizing the total cost may require a cost of defense that is higher than the cost of insult that induced the defense (Figure 4.4).

The structure of the costs of defense

The costs of defenses are of several types and can be measured in several ways. The evolutionary measure of a cost is a negative genetic covariance with some other fitness-increasing function. In model organisms, such negative genetic covariances are most convincingly measured as correlated

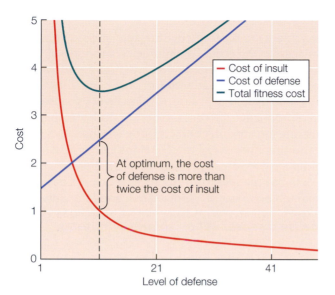

Figure 4.4 Total cost of defense can be minimal even when the cost of defense is greater than the cost of the insult. (After Nesse 2001.)

responses to artificial selection: one selects for improved function in one trait over several generations—usually at least 10, preferably 100 or more—and then measures the correlated responses in other traits. There are also short-term physiological measures of costs. All biological processes and structures have maintenance costs, and it takes energy, materials, and specific "infrastructure" to maintain defensive organs and tissues, such as the skin, the skull, the rib cage, and the lymphatic system. All induced responses have deployment costs that involve both energy and interference with other physiological functions. Some induced responses, such as immune responses, have additional costs; in this case, they are collateral tissue damage and other forms of immunopathology.

How defenses lead to disease

Defenses can either lead to disease because they fail to protect or because their costs are "excessive" from the point of view of patient and physician (perhaps not from the point of view of evolution). Whether the costs of defense are excessive depends on the relative priority of the defense being deployed (e.g., protecting from infection) versus the functions where the costs of the defense are being paid (e.g., digestion or respiration). Clearly, some of the costs, such as temporarily suppressed digestion, are more acceptable than others, such as temporarily suppressed respiration. In addition, costs that are acceptable in one environment or developmental stage can be excessive or purely detrimental in other environments, physiological states, or developmental stages. Thus, environmental change and the progress of the patient through her or his life history can shift the cost-benefit ratio and make costs that were acceptable elsewhere or earlier excessive here and now. That is one major reason for mismatches to modernity, which cause the diseases of civilization (see Chapter 8).

When do costs become unacceptable?

Because evolution gives priority to reproductive success, it buys improvements in reproductive success at the cost of maintenance and survival. Such costs, acceptable from an evolutionary perspective, are often excessive and unacceptable from the point of view of patients and physicians. Costs that were historically acceptable to evolution at a population level because they led to improvements in reproductive success are now unacceptable at the individual level and illustrate the striking conflicts between the interests of conscious individuals and the interests of unconscious genes. Examples include all cases of late-life disease whose prevalence and risk are increased by the antagonistically pleiotropic effects of genes that benefit early survival and reproduction at the expense of maintenance and later survival.

When do defenses cause disease? When should symptoms be treated?

That the costs of defense can be important causes of disease symptoms is particularly obvious in the case of infectious diseases. The most common cause of symptoms during infections is immunopathology, the damage caused by the immune response. Such symptoms include fever, anorexia, coughing, and diarrhea. All are either expressions of defense (e.g., coughing and diarrhea promote pathogen expulsion) or reflect the costs of defenses (e.g., diarrhea interferes with digestion).

Blocking symptoms that are caused by expression of defenses can temporarily alleviate suffering, but it can also interfere with the defenses. If the costs of defense are, in fact, excessive, blocking the symptoms can relieve suffering without negative consequences.

The big question is "When is it safe to interfere with symptoms that are caused by the expressions or the costs of defenses?" That question can only be properly answered by discovering the functions with which the defense is trading off. The problem is that the evolutionary logic of settling on a particular cost-benefit ratio is often obscure because it depends on environmental factors and physiological priorities that shaped the selection of defense mechanisms at various points in our evolutionary history.

■ SUMMARY

Environmental hostility is heterogeneous in space, in time, and over the life history of the organism. Defenses have costs. Because they participate in trade-offs, tweaking them has indirect consequences. Defenses allow disease either through failure to protect or through excessive costs. Costs that were evolutionarily acceptable at the population level may now be unacceptable to patients and physicians at the individual level. Some costs of defense are the main cause of disease symptoms.

Specialized defenses

When a hostile environmental factor is frequently encountered, it elicits the evolution of defense mechanisms that protect against its specific effects.

As with inducible maintenance programs, these defense mechanisms are usually modifications of homeostatic systems that are always operating, even when the environmental challenge is absent. These modifications often have specialized characteristics that are only expressed under conditions of environmental stress because, like all induced responses, they have costs, and it does not pay to deploy them unnecessarily. The environmental challenges that elicit such defense responses include starvation, dehydration, cold and hot temperatures, hypoxia, predators, injuries, toxins, and pathogens. The corresponding defenses include the fight-or-flight response, blood clotting and tissue repair, pain (nociception), inflammation, allergic defenses, detoxification, and immune responses to infection. All are specialized extensions of homeostatic and maintenance programs, as is particularly clear in the case of defenses against starvation, dehydration, and extreme heat and cold.

Starvation

We start with the normal function of the relevant homeostatic mechanism: Following meal consumption, pancreatic β-cells produce insulin in response to raising blood glucose level and as potentiated by gut hormones, the incretins. Insulin promotes glucose consumption in skeletal muscle and adipose tissue and promotes glycogen and lipid synthesis in the liver. During the fasting phase, blood glucose is maintained by reduced insulin secretion and increased production of glucagon by pancreatic α-cells. Combined reduction in insulin and increase in glucagon levels promotes glycogenolysis and gluconeogenesis, both contributing to sustaining blood glucose level. During prolonged fasting, glycogen stores are depleted, and gluconeogenesis becomes the main source of systemic glucose. Because major glucose-consuming tissues (skeletal muscle and fat) require insulin for glucose utilization, glucose allocation changes during fasting such that only the brain and other high-priority tissues continue consuming it in an insulin-independent fashion. Prolonged fasting also stimulates lipolysis, or mobilization of stores of triglycerides from the adipose tissue. Fatty acids then replace glucose as the main metabolic fuel for most tissues while the brain continues to use glucose. As fatty acid stores become depleted (in the early starvation phase), the liver starts synthesizing ketone bodies—acetone, acetoacetic acid, and β-hydroxybutyric acid—that become the fuel for most tissues, including the brain. When starvation becomes extreme, protein in skeletal muscle is broken down to provide amino acids as fuel sources.

Thus, metabolic defenses against extreme food deprivation are extensions of the fasting response that occurs normally between meals. Many additional metabolic and endocrine changes occur during starvation that shut down energy consumption by nonessential functions and promote energy generation from available metabolic fuels. Interestingly, some of these changes are orchestrated by stress hormones, glucocorticoids, and adrenaline that are also engaged under other types of environmental hostility.

Dehydration and volume depletion

Water and sodium (volume) balance are tightly regulated by several homeostatic circuits, including the renin-angiotensin system, aldosterone, natriuretic hormones, and vasopressin. Angiotensin-II (primarily from the liver) and aldosterone (from the adrenal cortex) are induced in response to loss of sodium (and therefore blood volume) and promote sodium reabsorption in the kidney, whereas vasopressin (from the neurohypophysis) is induced by water loss and promotes renal water retention and the feeling of thirst. These pathways operate to maintain normal water-sodium balance, but their activity is enhanced as dehydration becomes more severe. During dehydration, the sympathetic nervous system orchestrates additional adaptive changes in vascular tone and heart rate. All these adaptive responses are extensions of the normal homeostatic controls of electrolyte balance and blood volume. The limits of physiological adaptation to dehydration are largely dependent on the organism's urinary-concentrating ability, which is defined as the ratio of the urine to blood plasma osmotic pressure (the U/P ratio). In humans, the maximal U/P ratio is 4; in animals adapted to desert habitats, the U/P ratio can be as high as 14 for Mongolian gerbils and 26 for Australian desert-hopping mice, for example.

Cold and heat stress

In birds and mammals, core body temperature is held nearly constant. Changes in the ambient temperature stimulate adaptive thermogenesis or heat dissipation; that is normal homeostasis. When ambient temperature changes become extreme, the normal homeostatic mechanisms go into overdrive. Cold stimulates the hypothalamus to activate antagonistic skeletal muscles, inducing shivering thermogenesis. Heat can also be produced without shivering by uncoupled oxidative phosphorylation in brown fat, which is a major defense in neonates, infants, and young children. They have abundant brown fat near vital organs and near arteries that supply blood to vital organs, such as the brain. Prolonged cold exposure induces differentiation of beige adipocytes, which also generate heat by uncoupled oxidative phosphorylation. Cold adaptation reduces heat dissipation by decreasing cutaneous circulation and heat flow to the skin and extremities and by using countercurrent exchangers to retain heat in the body core. Heat adaptation increases heat dissipation by increasing blood flow to the skin and by activating sweat glands to increase evaporative heat loss. These induced responses are extensions of normal homeostatic controls of body temperature in mammals.

Hypoxia

Adequate oxygen is essential for survival because it is required for oxidative phosphorylation in the mitochondria. Tissues and organs vary in their sensitivity to oxygen deprivation: whereas fast twitch muscles use glycolysis to generate ATP and are thus relatively tolerant to short-term oxygen shortage, neurons and cardiomyocytes, which rely almost exclusively on oxidative phosphorylation, are exquisitely sensitive to oxygen deprivation.

That is why myocardial infarction and stroke are common consequences of local ischemia. In contrast to the starvation response, where stored triglycerides can be used as an energy source during food deprivation, there is little tolerance to hypoxia because no adequate tissue storage of oxygen exists in terrestrial animals like humans (aquatic mammals can store enough oxygen bound to myoglobin to allow for prolonged diving).

Local-tissue hypoxia can occur either physiologically (e.g., in exercising skeletal muscle) or pathologically (due to ischemia). Unlike water, nutrients, and temperature, however, environmental oxygen availability in terrestrial habitats does not vary over time. Oxygen tension does decline with altitude, though, and oxygen demand can transiently increase during exercise. The short-term response to hypoxia under these circumstances is hyperventilation and increased allocation of oxygenated blood to the brain and heart. The longer-term acclimation to reduced oxygen tension at high altitudes is mediated by increased production of erythropoietin by kidneys to increase red blood cell numbers. Both of these mechanisms are extensions of homeostatic control of oxygen level within an organism.

The human response to high altitude varies among populations: Andeans have higher hemoglobin concentration and percent oxygen saturation than do Tibetans at the same altitude, and both differ from people living at sea level, but Ethiopian highlanders do not differ in these two traits from people living at sea level (Beall 2006). An evolutionary response to high altitude is revealed in Tibetans, in whom alleles of a gene encoding the transcription factor EPAS1, which increases the production of red blood cells, are at significantly higher frequency than they are in Han Chinese (Beall et al. 2010). These different outcomes probably reflect in part the amount of time the different populations have spent at high altitude and the amount of gene flow they have experienced with lowland populations.

The fight-or-flight response

When a sudden, dramatic physical response is required for survival, such as during an attack by a predator or competitor or during injury from an accident, the fight-or-flight response is activated. The fight-or-flight response, also known as the generalized stress response, is controlled by the sympathetic nervous system and by the hypothalamic-pituitary-adrenal axis. The logic of the stress response is to mobilize resources (glucose, fatty acids, and oxygen) needed for high-priority organs (brain and skeletal muscle), while at the same time suppressing nonessential, lower-priority processes, such as digestion, anabolism, growth, and reproduction.

Within seconds of stress exposure, sympathetic output increases dramatically. That leads to increased heart rate and ventilation to increase the blood and oxygen supply to muscle and brain as well as glycogen breakdown and lipolysis to increase the availability of glucose and fatty acids to be used as fuels. At the same time, insulin production and digestion are suppressed, and glucagon release is increased. Catecholamines (epinephrine and norepinephrine) along with glucagon promote gluconeogenesis to sustain high blood glucose level. Within an hour of the stress response, the hypothalamus starts producing corticotropin-releasing hormone, which stimulates release of adrenocorticotropic hormone (ACTH) from the pituitary. ACTH induces glucocorticoid production by the adrenal gland. Glucocorticoids complement and potentiate the effects of catecholamines: they promote liver gluconeogenesis and fat and protein catabolism, enhance the vascular effects of sympathetic neurons, and induce insulin resistance. In addition, glucocorticoids inhibit hypothalamic production of growth hormone and gonadotropins, thereby suppressing hormonal pathways for growth and reproduction, which are costly and not essential in the hostile environments that elicit stress response. Thus, in many ways, the fight-or-flight response is an exaggerated version of normal sympathetic control of homeostasis.

Pain (nociception)

As a reaction to mechanical, chemical, and thermal damage, pain sensation is an essential, vital defense that helps reduce and avoid tissue damage. The lack of pain sensation (congenital analgesia) is life threatening and often lethal without special measures. Somatosensory neurons involved in pain sensation (nociceptors) can be activated by a variety of noxious stimuli, both exogenous and endogenous. The endogenous stimuli include signals released upon tissue damage, such as extracellular ATP, low pH, and bradykinin. Nociceptors can also be activated by noxious chemicals, heat, and cold. Although acute (primary) pain normally occurs only during tissue damage, secondary pain, which is mediated by unmyelinated slow-conducting C-fiber neurons, plays an important homeostatic role. Indeed, C-fiber nociceptors appear to be involved in monitoring tissue homeostasis by sensing local metabolites, pH, temperature, and other parameters in the local tissue milieu. By monitoring tissue "well-being," nociceptors elicit local and systemic responses that help maintain homeostasis. An exaggerated version of these responses elicits pain sensation and behavioral defense mechanisms, such as avoidance of painful stimuli.

Tissue repair and blood clotting

When cells are lost due to injury, they are replenished by proliferation and differentiation of the corresponding tissue stem cells (or in some cases, by mitosis of fully differentiated cells, such as hepatocytes). This tissue regenerative response is an enhanced version of normal tissue renewal, in which tissue stem cells continuously produce newly differentiated cells at rates that vary greatly among tissues. Tissue regeneration potential is typically related to tissue renewal capacity: tissues with high renewal rates, such as skin and intestinal epithelia, have high regenerative capacity, whereas tissues with low or no renewal capacity, such as neurons and cardiomyocytes, have low or no regenerative capacity and are therefore much more vulnerable to damage.

Tissue regeneration (replacement of lost cells) is a component of a broader program of tissue repair, which, in addition to cell replenishment, involves remodeling of the extracellular matrix and angiogenesis and is usually accompanied by inflammation. Tissue renewal and repair correspond to constitutive and inducible maintenance programs, respectively. They are essential for long-lived animals, and they decline with age. They also create a crucial vulnerability for cancer development at old age (see Chapter 6).

When there is vascular damage, the blood-clotting cascade is activated to prevent blood loss and to promote repair of the damaged blood vessels. In addition to its protective role in damaged blood vessels, the clotting cascade can be induced by inflammatory signals to prevent dissemination of pathogens. Although blood clotting per se normally occurs only under conditions of vascular damage or inflammation, it is an extension of the hemostasis that operates normally to promote vascular integrity because blood vessels cannot operate normally without platelets and other components of the clotting machinery.

Inflammatory response

Severe perturbations of tissue homeostasis, such as during infection or injury, elicit the inflammatory response. Depending on the nature and the extent of the insult, the response can be either localized or systemic. Inflammation is initiated by tissue resident cells, most commonly macrophages and mast cells, which are equipped with receptors to detect pathogens and damage-associated signals. Once activated by the appropriate stimuli, these cells produce inflammatory mediators: cytokines, chemokines, biogenic amines, leukotrienes, and prostaglandins. In addition, C-fiber nociceptors monitor the tissue microenvironment and produce neuropeptides to elicit a local inflammatory response to tissue injury. Collectively, the inflammatory signals initiate local vascular response, including vasodilation, edema, and leukocyte recruitment to the site of infection or injury. Formation of the inflammatory exudate (plasma and leukocytes) is responsible for the cardinal signs of inflammation (rubor, tumor, calor, dolor, i.e., redness, swelling, heat, and pain). The goals of the local inflammatory response are to eliminate or neutralize the noxious stimulus, to repair the damaged tissue, and to protect the rest of the organism from the deleterious consequences of the noxious agent.

The systemic inflammatory response is orchestrated by circulating inflammatory cytokines. In the liver, these cytokines induce the acute phase response: a dramatic change in hepatic protein secretion into the circulation. Proteins involved in host defense, blood clotting, and tissue repair are increased, while the production of proteins involved in normal homeostatic functions (e.g., serum albumin) is suppressed. Inflammatory mediators, especially prostaglandins, can also act on hypothalamic neurons to alter homeostatic set points for body temperature and appetite. The resulting fever and anorexia along with fatigue, social withdrawal, sleepiness, and suppressed sex drive constitute the so-called sickness behaviors associated with acute illness, including infectious diseases. Interestingly, many of the changes induced by acute inflammation are the opposite of those induced during the fight-or-flight response. There are, however, important overlaps in the two defense reactions. As with the glucocorticoids, inflammatory mediators suppress hormonal pathways involved in the control of growth and reproduction. Also, as with the fight-or-flight response, inflammatory signals override some homeostatic controls to meet the demands of the higher physiological priorities of surviving infection or injury.

Inflammation is a costly defense. The acute inflammatory response can cause tissue damage, while chronic inflammation can lead to a wide variety of diseases of homeostasis. As will be discussed in Chapter 8, high-benefit, high-cost traits are particularly vulnerable to mismatches with the environment. Many chronic inflammatory diseases may be a consequence of such mismatches.

Allergic defenses

Allergic defenses operate in barrier tissues, which include skin and the mucosa of the upper respiratory tract, the gastrointestinal (GI) tract, and the urogenital tract. They are induced responses to noxious substances, such as venoms, irritants, and xenobiotics that include phytochemicals, various enzymes, and even some heavy metals. These noxious substances are collectively known as allergens. Some allergens are innocuous but can elicit allergic reactions if they are mistakenly associated with noxious activity by the immune system. The main purposes of allergic defenses are to expel noxious substances and to reinforce the barriers that prevent their entry. All symptoms of allergy are manifestations of these defenses: mucus production to enhance mucosal barriers (rhinorrhea, or runny nose), airway obstruction, bronchoconstriction (to prevent further exposure to airborne substances), sneezing and coughing (to expel noxious substances from the airways), vomiting, peristalsis and diarrhea (to remove noxious substances from the GI tract), and itching (to remove noxious substances from the skin). Although these symptoms of allergic reactions are usually thought of as pathological, they are, in fact, defenses and are normally protective. These defenses do come with significant costs, including discomfort, but produce benefits analogous to those of pain: some of them—for example, itching and nausea—are meant to be unpleasant to induce behavioral avoidance of noxious environmental agents.

The detoxification response

Exposure to toxic environmental chemicals induces a different kind of protective response, one based on metabolic modification of the noxious chemicals and their subsequent excretion. Noxious xenobiotics have chemical features that make them incompatible or outright toxic to the organism's biochemistry. For example, hydrophobic xenobiotics exert toxic effects when they accumulate inside the cells and interact with cellular membranes and proteins, causing their damage. Similarly, electrophilic compounds can form adducts with proteins, nucleic acids, and membrane lipids, resulting in their damage, aggregation, misfolding (proteins), peroxidation (membrane lipids) and mutations (DNA).

Several lines of defense against noxious chemicals evolved to protect against their toxic effects. The best known is the detoxification response mediated by ligand-activated transcription factors, including the aryl hydrocarbon receptor (AhR), the pregnane X receptor (PXR), and the constitutive androstane receptor (CAR). These receptors are activated by a diverse array of xenobiotics, including dioxin, phenobarbitals, and many phytochemicals. When activated by their xenobiotic ligands, these receptors induce the expression of genes involved in detoxification and excretion of the xenobiotics. These genes fall into three functional categories corresponding to the three steps in the detoxification response. Phase I genes encode members of the cytochrome P450 family. These enzymes oxidize hydrophobic chemicals, typically generating a hydroxyl group. They can be further modified by phase II enzymes that introduce additional charged hydrophilic groups, such as sulfate or glucoronic acid. The addition of these groups makes the target chemicals water soluble, facilitating their excretion. Finally, phase III of the detoxification response is mediated by membrane transporters and pumps that export noxious chemicals (often after their modification by phase I and phase II enzymes) for excretion in the urine. Phase II and phase III detoxification can also operate independently of CYP450 enzymes. For example, N-acetyl transferases participate in detoxification of acrylamines, and glutathione transferases play essential roles in the detoxification of reactive oxygen species.

This ancient defense system, which protects against the harmful effects of reactive chemicals, presumably evolved as a defense against the secondary metabolites that plants produce to defend themselves from herbivores. In the modern world, the detoxification system can provide protection against some artificial xenobiotics, such as industrial pollutants, and toxic drug metabolites. In addition to defending against environmental xenobiotics, the detoxification pathways also play an important role in neutralizing the toxic by-products of normal metabolism. These by-products are sometimes referred to as endobiotics to highlight their endogenous origin. For example, AhR, CAR and PXR, in addition to detecting xenobiotics, also monitor endogenous metabolites and induce their detoxification and excretion. One example of an endobiotic by-product of metabolism is methylglyoxal (pyruvaldehyde), a toxic by-product of glycolysis and threonine catabolism. It is detoxified by the glyoxalase system to prevent the deleterious consequences of aldehyde group reactivity. Detoxification of endogenous by-products of metabolism is a constitutive maintenance program that is conceptually distinct from defense against environmental insults, although the two are clearly mechanistically and evolutionarily related.

The immune response to infection

The immune system defends against microbial infections and plays an essential role in the survival of multicellular organisms. In humans and other vertebrates, the immune system has two components, innate and adaptive. The innate immune system detects infection by recognizing conserved microbial products, such as lipopolysaccharides and peptidoglycans. These structures are unique products of microbial metabolism that have essential housekeeping functions that the pathogens cannot do without.

These properties both permit discrimination between microbial and self-molecules and ensure that pathogens do not evolve escape mutants because any such mutants would be crippled by the loss of essential molecular components of microbial physiology. The innate immune system can also detect infection indirectly by sensing the stereotypical activities of virulence factors, toxins, and allergens. For example, many Gram-positive pathogens produce secreted pore-forming exotoxins, which cause, among other things, potassium efflux from the host cell. A cytosolic sensor called the NLRP3 inflammasome can detect this loss of membrane integrity and induces an inflammatory response against the pathogen. In this case, the characteristic biochemical activity of the exotoxin is detected by the innate immune system rather than the toxin itself. A similar principle is involved in sensing allergens, which often have enzymatic (e.g., proteolytic) activities. The innate immune system can detect these biochemical activities as a proxy of allergen exposure. The detection of virulence activities also helps distinguish between commensal microbes and pathogens.

Upon detection of infection by one of these mechanisms, the innate immune system activates the inflammatory response and other antimicrobial mechanisms, including phagocytosis, the complement system, and the secretion of antibacterial defensins or the induction of antiviral genes. This rapid response is essential to contain an infection of rapidly replicating pathogens and, in some cases, is sufficient to control infection without the involvement of adaptive immunity.

The innate immune system also activates the adaptive immune system, which is mediated by T and B lymphocytes. They recognize antigens with antibodies (Figure 4.5) and T-cell receptors that are generated at random by somatic recombination, gene conversion, and random pairing of receptor subunits. These processes generate a set of millions of antigen receptors that can recognize almost any antigen. Lymphocytes with antigen receptors that are specific for microbial antigens are selected to proliferate to generate clones of effector cells as well as long-lived memory cells.

Because the antigen receptors are initially formed by random processes, they cannot determine whether the antigens for which they are specific were produced by the host, by microbial pathogens, or by innocuous environmental substances. Activation of lymphocytes therefore needs to be controlled so that they do not react inappropriately to self-antigens or to innocuous environmental antigens.

It is the innate immune system that determines the origin of an antigen: whether it is from a virus, from an extra- or intracellular bacterium, or from a eukaryotic parasite. That is why it is required to activate the appropriate adaptive immune responses, which differ with the type of pathogen detected.

A key feature of adaptive immunity is memory. Some of the T and B lymphocytes that are specific to a pathogen form a special pool of long-lived memory cells that can be rapidly reactivated if the pathogen is encountered again. This property of the adaptive immune system is the basis of

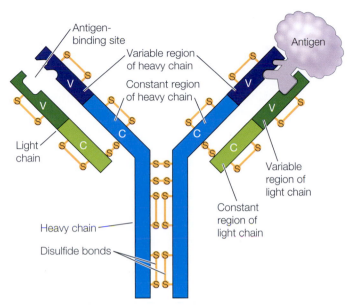

Figure 4.5 Antibodies are formed by combining the protein products of many polymorphic genes in gene families. V, variable region; C, constant region.

vaccination. Vaccines mimic infection to produce immunological memory that protects against real future infections.

Diseases of exaggerated defense

Each of the defenses described above is a benefit associated with costs. Those costs are expressed when the defense is exaggerated or otherwise inappropriately deployed, and the greater the benefit potentially conferred by a defense, the greater the cost that evolution will tolerate. Defenses do not need to be exaggerated to produce deleterious effects that can be viewed as symptoms of a disease. In many cases, however, expression of defenses can be excessive in intensity or duration, resulting in pathological consequences that range from discomfort (e.g., itching) to mortality (e.g., septic shock). Although death is clearly a maladaptive behavior of the defense system, it is hard, in most cases, to draw a line between evolutionarily justified costs of defense and truly pathological exaggeration of a normally adaptive defensive response. A large group of human diseases are nevertheless a consequence of defenses that are exaggerated, mistargeted, or activated inappropriately (at a wrong time or place). Examples of the corresponding defenses of such diseases are immune defenses, coagulation and thrombosis, tissue repair, fuel allocation and defense from starvation, defenses from dehydration and volume depletion, avoidance of pathogens, and predator avoidance behavior, each discussed below.

IMMUNE DEFENSES   When exaggerated or mistargeted, immune defenses can lead to autoimmunity, allergy, asthma, septic shock, and other immunopathologies. Although there is a strong genetic component in pre-disposition to autoimmune diseases, asthma, and allergies, the incidence of these diseases increased dramatically in the second half of the 20th century, and continues to rise, suggesting an important contribution of environ-mental factors (see Chapters 3 and 8). Conditions like septic shock and anaphylaxis may reflect imperfections of the control mechanisms. Multiple feed-forward loops combined with amplifying positive feedback mecha-nisms are designed to provide a rapid emergency response to overwhelm-ing infection (sepsis) or to exposure to venoms and toxins (anaphylaxis). These features, however, also create vulnerability to runaway processes that can quickly become unmanageable. Thus, septic shock is characterized by cascading failures of multiple organs. Similar to cascading failures of power grids leading to massive blackouts, it is a common vulnerability of highly connected systems in which failure of one component increases the strain on, and chances of failure of, the remaining components.

COAGULATION AND THROMBOSIS   Essential defenses against vascular injury are coagulation and thrombosis, but they come with a risk of embo-lism, a blockade of blood vessels by a thrombotic clot that breaks out from the initial location and lodges elsewhere in the circulation, causing local obstruction of blood supply. Such an obstruction can lead to tissue anoxia with consequent ischemia and infarction. The brain and the heart are at particular risk of embolism because neurons and cardiomyocytes are exquisitely sensitive to oxygen deprivation. Pulmonary embolism is another life-threatening condition because it interferes with respiratory function, leading to systemic anoxia. From the evolutionary perspective, the vulner-ability of this defense system to pathology is due in part to the following design feature: to be effective, blood clotting has to operate extremely fast to prevent hemorrhage. This speed is achieved by engagement of positive feedback mechanisms that amplify the response. Positive feedback loops create vulnerabilities of two types: they can lead to excessive responses, and they can lead to vicious cycles. Both of these vulnerabilities are pres-ent in the blood-clotting system. Thus, inflammation promotes the clotting response, whereas embolic obstruction promotes local inflammation. It is another example of a cascading failure that makes embolic conditions particularly challenging to treat.

TISSUE REPAIR   When exaggerated, tissue repair can lead to two types of pathological sequela: fibrosis and neoplasia. The tissue repair response has two components: regeneration (replacement of loss cells) and rebuilding of the lost extracellular matrix, including basement membranes. Exces-sive regeneration can produce tissue overgrowth, which in the context of oncogenic mutations is a critical contributor to tumor growth. When tissue injury involves damage to the basement membrane and other extracellular matrix components, regeneration is not sufficient to repair the tissue, and new matrix production is necessary. For reasons that remain unknown, this process operates perfectly in young children until about 5 years of age. At later ages, it is accompanied by fibrosis, resulting in scarring. Fibrosis also accompanies the repair process in tissues in which regeneration is impossible or ineffective or when cell loss is too extensive to be handled by regeneration. In these cases, the proliferating fibroblasts and the collagen produced by them fill in the "gaps" in tissues, resulting in fibrotic degeneration. Fibrosis may be a lesser evil than net loss of tissue mass, but when excessive, it results in pathological complications that include loss of tissue function (e.g., liver cirrhosis and pulmonary sarcoidosis).

FUEL REALLOCATION AND DEFENSE FROM STARVATION   The essential metabolic defenses, fuel reallocation and defense from starvation, are good examples of defenses with costs that never had to be paid until modern times. The ability to store excess calories as fat must have been life saving in most of human evolutionary history, but the inherent vulnerability to obesity and its complications has only emerged in modern times of continuous and unlimited food supply. Fuel reallocation is important to meet changing physiological priorities, such as during an infection, but in the context of obesity and chronic inflammation, this very feature can lead to the development of type 2 diabetes.

DEFENSES FROM DEHYDRATION AND VOLUME DEPLETION   The defenses from dehydration and volume depletion are based on water retention and sodium retention, respectively. These processes are tightly controlled by several endocrine mechanisms that evolved to operate in environments chronically deficient in sodium chloride. Salt consumption has increased dramatically in modern times, leading to a significant strain on sodium homeostasis. Because sodium plays a critical role in blood volume control, one consequence of excessive salt intake is higher incidence of hypertension.

AVOIDANCE OF PATHOGENS   A behavioral component of defense from infections is avoidance of pathogens. This defense is based on the detection of olfactory, gustatory, or visual cues that function as proxies of high microbial density. The smell or sight of rotten food and feces is repulsive to humans, as is the sight of humans suffering from contagious diseases. These avoidance mechanisms are critical for reducing the exposure to pathogens, but when excessive, they can interfere with normal behaviors. The maladaptive form of avoidance behavior, known as germophobia, is a common type of obsessive-compulsive disorder.

PREDATOR AVOIDANCE BEHAVIOR   Like pathogen avoidance, predator avoidance behavior is based in part on the detection of cues associated with a high risk of predation, such as open spaces and unfamiliar surroundings. Excessive expression of this behavioral defense can lead to maladaptive behavioral disorders, including agoraphobia.

■ SUMMARY

Organisms respond to environmental stressors with physiological adaptations that extend normal homeostatic mechanisms. Although at the extremes these responses have some specialized characteristics, those specialized characteristics are evolutionary modifications of processes that operate in the absence of the stressors. That may seem less obvious for immune defense, the detoxification response, and tissue repair than it is for starvation, dehydration, and thermoregulation. In all these cases, however, there is a nondefensive counterpart. The immune system manages the microbiota and tissue homeostasis, the detoxification system handles by-products of metabolism, and tissue repair is an extension of normal tissue renewal. Evolution created defenses out of preexisting mechanisms that had evolved for normal, nondefensive purposes. The exaggeration of each type of defense produces a characteristic associated disease.

Key characteristics of defenses

Two related dichotomies frame the key characteristics of defenses. One is constitutive versus inducible, and the other is innate versus acquired. While not absolute, these distinctions highlight important features of defense mechanisms.

Constitutive versus inducible

Whether a defense mechanism is constitutive or inducible depends on its costs, its deployment requirements, and whether the environmental challenges that shaped its evolution are present always, occasionally, or periodically. Most defenses are inducible on demand because they are costly, unnecessary, and, in the absence of a threat, often detrimental. They are an excellent example of traits that satisfy the functional criterion for recognizing adaptations. The costs of much of the immune response— energy and immunopathology—are too high to allow constitutive activation in the absence of infection. It does, however, have low-cost components, including secreted defensins and immunoglobulin A, that are constitutive because they are continuously needed to maintain barrier immunity and because they do not damage tissues.

When several defense mechanisms protect against a given challenge, they are usually induced in order of increasing costs. For example, avoidance is cheaper than repair. Anticipatory responses can be induced when challenges occur with predictable periodicity or when a proxy signal is detected, indicating that they are about to occur. For example, undernourishment in the womb and in infants induces a state that alters metabolism for years, protecting the developing brain but increasing risks of late-life diabetes, obesity, and cardiovascular disease in the so-called metabolic syndrome.

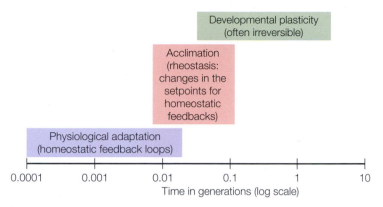

Figure 4.6 Responses are inducible on different time scales. (After Gluckman pers. comm. 2011)

Innate versus acquired

Innate defenses can be adjusted to handle environmental challenges by individual experience. A second exposure to a noxious substance can result in a more rapid and efficient response to subsequent exposures. An example of this process, known as *hormesis*, is the response to repeated exposure of low levels of a toxin that would be fatal at high dose. Repeated exposure to low doses induces expression of detoxification genes and other adaptations that allow a subsequent lethal dose to be tolerated. That is how, for example, Edmund Dantès in *The Count of Monte Cristo* protects Valentine de Villefort from poisoning.

Hormesis is related to acclimation, although the two often operate on different timescales. It is distinguished from acclimation by referring to repeated discrete environmental exposures, whereas acclimation is usually used to describe responses to continuous exposures to changes in factors such as temperature, humidity, and oxygen tension.

Time scales of deployment

Inducible defenses are reversible and are deployed on different time scales (Figure 4.6). Rapid physiological adaptation—hyperventilation, shivering, increased heart rate—happens in seconds to minutes. Lipolysis, gluconeogenesis, and the immune response react on a scale of hours to days. Acclimation takes days to weeks to months to years, depending on the environmental factor. Developmental plasticity is an extreme and often irreversible response that is often induced at an early stage of development and lasts a lifetime.

Similarities in immune and behavioral defenses

Defenses mediated by the immune and nervous systems both have two distinct components: innate and acquired (or learned). Innate components are genetically specified, hard-wired programs that are activated by a limited number of biologically salient cues, such as bacterial lipopolysaccharides or the smell of a predator. Acquired or learned defenses can be activated by an almost unlimited number of neutral cues through their association with the biologically salient clues that trigger innate responses. Thus, antigens by themselves are neutral unless they are recognized by T and B lymphocytes in the context of innate immune stimuli. Similarly, most environmental clues perceived by the sensory system are neutral unless coupled with biologically salient signals associated with food, predators, mates, and so forth. Acquired systems are very flexible and can be adjusted to unpredictable environments and associated hostilities, such as pathogens, predators, toxic plants, and venomous animals. The acquired components of both immune and nervous systems also have the benefit of memory, which allows for a more rapid and efficient response upon a subsequent encounter with the same cue or a challenge. The immune and nervous systems thus share important design features. They are also capable of communicating with each other, sharing information on the environment (Loisel, Alberts, and Ober 2008).

The benefits and costs of flexibility and homeostasis

At first sight, the comparison of invertebrates, which tend to rely on innate immunity and behavior for defense, with vertebrates, which use both innate and acquired immunity and innate and learned behavior, would suggest that vertebrates have impressive advantages. That conclusion, however, is not entirely clear because the adaptive systems used by vertebrates come with significant vulnerabilities. The adaptive immune system creates opportunities for autoimmune and allergic diseases that are absent in invertebrates, and the vertebrate capacity for learning is associated with mental diseases and behavioral abnormalities, including obsessive behavior, paranoia, and phobias. The costs of flexibility thus include risks of new types of disease.

Redundancy, compensation, and compatibility of defenses

Most defenses have multiple components that act in concert to provide optimal protection. This coordination is most obvious in the case of immune defenses, when multiple effector mechanisms (antibodies, the complement system, phagocytes, cytotoxic lymphocytes, etc.) operate to deal with any given pathogen. These defense pathways can compensate for one another to some degree when one of the pathways is inactivated by mutations or by pathogen evasion mechanisms. Such compensation makes the system more robust and efficient, although the existence of multiple compensatory mechanisms is often confused with redundancy. Another important advantage of having multiple complementary defense mechanisms is that it helps minimize the costs of defense: each defense component operates within a characteristic dynamic range, with the higher degree of expression being associated with higher costs (resulting in immunopathology). Engagement of multiple defense mechanisms, however, permits cost allocation between them so that each does not need to be expressed at the higher end of the dynamic range. Consequently, the cost of defense is reduced without compromising its effectiveness.

Interestingly, when one of the mechanisms of defense is compromised (e.g., due to mutation or evasion by the pathogen), the remaining functioning pathways have to operate at higher magnitude or duration, often resulting in immunopathology. That is why immunodeficiency can paradoxically lead to autoimmunity. Another important characteristic of defenses is their compatibility: some defenses are compatible in that they neither cooperate nor interfere with one another; other defenses can be synergistic in that they cooperate with one another; and, finally, some defenses are incompatible in that engagement of one defense response interferes with another, thereby creating crucial vulnerabilities. Thus, the inflammatory response is incompatible with the detoxification responses, and cold adaptation and immune defense are both incompatible with the starvation response. The causes of incompatibilities can be due either to the use of shared limited resources (e.g., energy) or to functional incompatibility. An example of the latter is the immune response against viral and bacterial pathogens: they can be functionally incompatible, leading, for example, to secondary bacterial pneumonia following flu infection.

■ SUMMARY

Defenses are inducible when costly, and redundant defenses are induced in order of increasing cost. Hormesis occurs when a first exposure to a threat induces defenses that make later exposures less harmful; its existence has long been known to those operating in environments in which poisons are frequently encountered. Induction of defenses takes time. Some are fast and react in seconds; others take weeks, months, or even years to complete. Adaptive and acquired defenses have the benefits of flexibility and memory but bring with them vulnerabilities to new types of diseases.

Defense strategies: Avoidance, resistance, and tolerance

The three major types of defense strategies are avoidance, resistance, and tolerance. Reliance on a particular defense strategy depends on the nature of the threat. Thus, avoidance is the only viable strategy when dealing with most predators; resistance is important for defense against invasive pathogens; and tolerance is the only option when dealing with cold, starvation, and dehydration. All three strategies are used in defense against infections. We use this defense as an example to illustrate their distinct roles.

Avoidance behaviors, which reduce exposure to pathogens, include both innate and learned aversion to signals associated with high microbial density. These signals are detected primarily through olfaction and taste, although visual cues can also play a significant role in humans. Nausea and feelings of disgust help prevent infections and exposure to bacterial toxins and noxious chemicals. Avoidance is a normal defensive response, but it also has costs. Extreme responses like germophobia, which is one type of obsessive-compulsive disorder, are exaggerated maladaptive, costly responses to nonexistent threats.

Selection designed resistance mechanisms to eliminate pathogens once they invaded the host. Our principal resistance mechanism is based on the immune system, which reduces pathogen burden through detection, destruction, and elimination. Resistance clearly has advantages, but it also has significant costs. Most types of immune response carry the risk of immunopathologies, including autoimmune diseases, allergies, and asthma. There thus comes a point at which tolerating a disease is less costly than resisting it.

Tolerance mechanisms, which reduce the negative effects on host fitness of a given level of pathogen burden, operate through tissue protection and repair and other forms of *inducible maintenance*. Because tolerance mechanisms do not affect pathogens directly, they impose less selection on pathogens than do resistance mechanisms; in other words, they provoke less evolutionary response. Consequently, there is less of an arms race, and in sharp contrast to resistance genes—like the defensins and major histocompatibility complex (MHC) families—there is less variation among individuals for tolerance genes. Tolerance mechanisms are part of the toolkit used by the immune system to manage homeostatic relations with the microbiota. Tolerance can also have costs; for example, tissue protection may interfere with normal tissue function. Thus, squamous metaplasia of the epithelium makes it more resistant to mechanical stress, but it comes at a price of reduced functionality, such as gas exchange or nutrient transport.

Conceptualizing resistance and tolerance

Resistance and tolerance can be conceptualized as reaction norms of fitness to pathogen burden (Råberg, Sim, and Read 2007). A reaction norm is a range of phenotypes produced by the same genotype as a function of the environmental variation. A genotype is more resistant if it can better reduce pathogen burden (Figure 4.7A) and is more tolerant if it can better maintain fitness with a given level of pathogen burden (Figure 4.7B).

Defense and physiological priorities

All defenses are designed to promote survival in the face of hostile environmental factors. Organismal survival, in turn, depends on the continuous performance of vital physiological processes. Consequently, an important defense strategy is to insulate the vital organs, such as the brain and heart, and the vital processes, such as blood circulation and respiration, from environmental challenges as well as from the costs of defense. It can be achieved in several ways.

First, the resources that become limited as a result of environmental challenge are allocated according to the physiological priorities of the functions

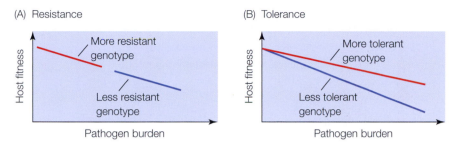

(A) Resistance

(B) Tolerance

Figure 4.7 (A) Resistance and (B) tolerance conceptualized as reaction norms. (After Råberg, Sim, and Read 2007.)

performed by different tissues and organs. For example, during prolonged fasting, the brain continues to use glucose at the expense of other tissues because glucose uptake in the brain does not depend on insulin, unlike skeletal muscle and fat. Similarly, during hypoxia the better-oxygenated blood is disproportionately distributed to the brain, as is warmer blood during hypothermia.

Second, vital tissues and organs are better protected than non-vital organs from mechanical, chemical, and inflammatory damage, so their performance is relatively less affected by environmental insults. Thus, the brain, heart, and lungs are protected from mechanical damage by the skull and rib cage. In addition, the brain and gonads are protected from toxic chemicals by the blood-brain and blood-gonad barriers.

Finally, certain tissues and organs known as immune privileged sites, including the brain, the gonads, and the eyes, do not permit immune responses that can cause tissue damage. This fact illustrates that the cost of defense can be differentially allocated between tissues so that the most vulnerable tissue, or the tissues and organs performing the higher priority functions, pay less of the cost.

Tissue vulnerability to damage (whether environmental or inflicted by the defense response) depends on several characteristics, including intrinsic damage susceptibility, renewal and repair capacity, functional autonomy, and damage sequela (Figure 4.8). These characteristics define a tissue's tolerance capacity, which in turn determines priorities in terms of cost allocation.

Thus, the decision chain in defense can be summarized as follows: if possible, avoid the threat by using vision, taste, and smell; if infected, choose either to resist or to tolerate, making that decision based on the costs and benefits of resistance, which vary among tissues and organs. To minimize the costs of inflammation, critical organs have immune privilege.

Failure of tolerance can contribute to infectious disease mortality

The issue of whether to resist or to tolerate an infection was strikingly raised by the high costs of resistance displayed in the 1918 influenza pandemic. Lasting from January 1918 to December 1920, it infected about 500 million

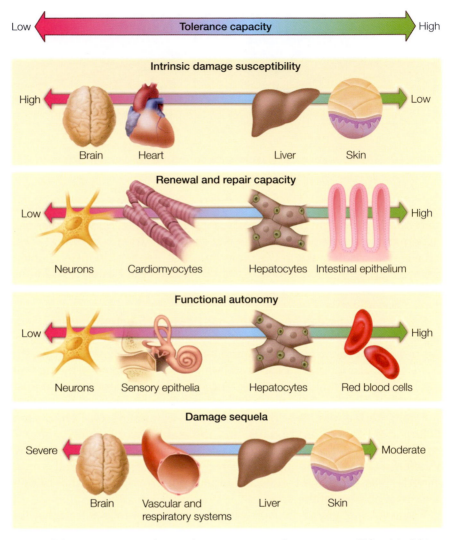

Figure 4.8 Capacity to tolerate damage varies with tissue type. (After Medzhitov, Schneider, and Soares 2012.)

and killed 50–130 million people, or about 3% of the global human population. Most victims were healthy, young adults with strong immune responses (Figure 4.9). The epidemic was caused by an H1N1 strain of the influenza virus, which is named according to the variants of the hemagglutinin and neuraminidase genes, whose products are displayed on its surface and interact with the host immune system. The 1918 strain of the virus is thought to have elicited an intense inflammatory response in the lung epithelium, a "cytokine storm" that was followed by the secondary bacterial infections that caused the pneumonia from which most people actually died. This response is an example both of incompatible defenses (against respiratory viral and

(A) Influenza case rates

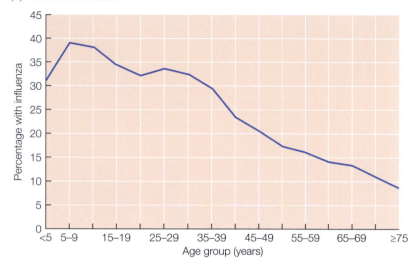

(B) Pneumonia rates

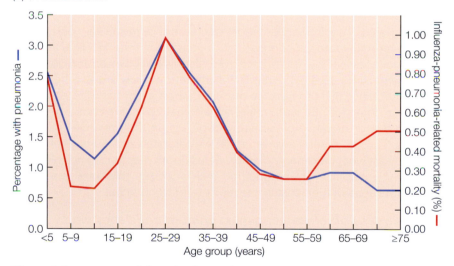

Figure 4.9 (A) Rates of clinical influenza by age group and (B) rates of pneumonia and mortality due to pneumonia by age group in the 1918 influenza epidemic. (After Brundage and Shanks 2008.)

bacterial infections) and of the failure to tolerate inflammatory damage in a vital organ, thus interfering with a key function and resulting in mortality.

It is not just acute inflammation that is costly. The inflammation caused by infectious diseases in childhood may increase the risk of noninfectious diseases in adults, as suggested by the correlation between the decrease in incidence of childhood diseases in Sweden and the decrease in deaths of adults (Figure 4.10; Finch and Crimmins 2004). Between 1751 and 1940, the

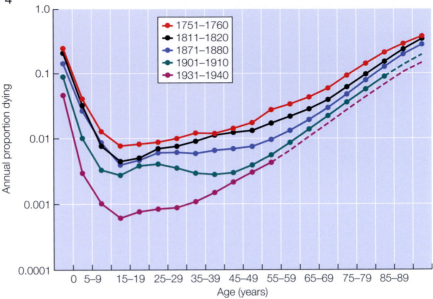

Figure 4.10   Cohort mortality in Sweden, 1751–1940. Dotted lines represent projections. (After Finch and Crimmins 2004.)

mortality rates of adult Swedes steadily decreased. By the 1930s, 65-year-olds had the mortality rate experienced by 45-year-olds in the 1750s. Other evidence indicates that the risks of cardiovascular disease, cancer, and dementia all rise with cumulative exposure to inflammation. They are another major cost of resistance that prompts the question: when is it better to tolerate than to resist an infection? Because tolerance also has costs, the choice between resistance and tolerance presumably depends on their relative costs and benefits. Thus, a general answer is that we tolerate what we cannot resist, and we resist what we cannot tolerate; we use both resistance and tolerance whenever possible; and when neither resistance nor tolerance is a viable option (because of their costs), infections tend to be lethal.

A classic case of tolerance: Simian immunodeficiency virus in its natural hosts

The closest relative of the human immunodeficiency virus (HIV) that causes AIDS in humans is the simian immunodeficiency virus (SIV) found in nonhuman primates in Africa, including African green monkeys and sooty mangabeys. The virus replicates in those hosts to produce viral loads similar to those in humans with HIV, but the hosts do not develop a disease similar to AIDS. Instead, they manage to tolerate the infection (cf. Figure 4.7B). Several mechanisms in the host are thought to be involved (Figure 4.11), including more rapid resolution of the innate immune response, better maintenance of mucosal barriers, and less infection of key immune cells. All these mechanisms reduce the level of chronic immune activation, which means less immunopathology, better homeostasis, and no progression to AIDS (Chahroudi et al. 2012). In addition, SIV has evolved to produce accessory immunomodulatory proteins that reduce pathogenesis in the natural hosts.

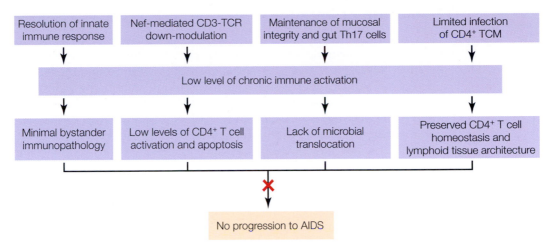

Figure 4.11 Mechanisms responsible for low immune activation and lack of disease progression in SIV-infected natural hosts. CD3, cluster of differentiation 3; TCR, T-cell receptor; CD4+ TCM, T cells expressing cluster of differentiation 4 glycoprotein; Th17, T helper 17 cells producing interleukin 17. (After Chahroudi et al. 2012.)

Tolerance has its own costs

Like resistance, tolerance has it own costs. The mechanisms that promote tolerance often do so at the expense of other functions; thus, stress adaptation and cytoprotection can interfere with cellular and tissue functions. For example, quiescent cells are more resistant to cell death than are dividing cells, but the protection comes at the expense of proliferation; likewise, squamous epithelium can better tolerate mechanical damage than columnar epithelium, but at the expense of absorptive function. In addition, there can be indirect costs of relying on tolerance in that tolerance is defined as a reduction of the negative impact of an infection on host fitness. Whenever the negative impact is not completely eliminated, the remaining impact is, in effect, the cost of relying on tolerance. For example, many animals can tolerate parasitic infections to some degree, but they still suffer significant negative consequences to their health.

■ SUMMARY

Avoidance, resistance, and tolerance are distinct but complementary host defense strategies. Depending on the environmental threat, be it a pathogen or a predator, only some defense strategies are feasible options. Avoidance and tolerance are the best choices when resistance is futile or too costly; resistance is the best choice when avoidance is impossible and tolerance is insufficient. In most cases, however, a combination of these strategies is used to achieve the optimal defense. Some infectious diseases are consequences of failed avoidance, some result from failed or excessive resistance, and some are best explained by insufficient tolerance.

Evolution of immunity

One major function of the immune system is to protect the host organism against pathogens; another equally important function is to manage healthy relations with the commensal microbiota. Performing these functions poses significant challenges for the immune system: the pathogens are highly diverse, they evolve much faster than the host, and they often need to be destroyed and eliminated while sparing the host tissues and commensal microbiota. The traits that evolve to enable these functions have high benefit (they promote survival) and high cost (they create vulnerabilities to many diseases). Although some principles of immune system function appear universal, their specific implementations vary among animal groups, reflecting different anatomical and physiological constraints, life history traits, and ecology, among other factors.

The vertebrate immune system has two components. The innate immune system, which shares many elements with invertebrates, provides an immediate, rapid, and more general response. The adaptive immune system, which is unique to vertebrates, takes longer to respond and provides both memory and greatly enhanced flexibility and precision.

The innate immune system

The innate immune system is found in all animals. It consists of several functionally autonomous defense modules that evolved to deal with different classes of pathogens, in different tissues, using different defense mechanisms.

Surface epithelia—skin and mucosal epithelia—provide an essential barrier defense that keeps most microorganisms away from the internal compartments. Although skin forms an efficient mechanical barrier to pathogens, the mucosal epithelia of the gastrointestinal, respiratory, and urogenital tracts, in addition to being physical barriers, also produce mucins and a variety of antimicrobial peptides that are embedded in the mucous layer. Together, these defenses form one of the most important modules of host defense that appears to be universal in the animal kingdom.

Phagocytes (macrophages and neutrophils) are essential for defense against bacterial and fungal pathogens. They are equipped with several potent antimicrobial mechanisms, including lysozymes that digest bacterial peptidoglycan (a polymer that forms bacterial cell walls); phagocyte oxidase and nitric oxide synthase, which produce reactive oxygen species (ROS) and nitric oxide, respectively, both of which are highly toxic to microbial cells; and a variety of proteases and hydrolytic enzymes that kill phagocytosed and extracellular pathogens. Macrophages and neutrophils phagocytose microbial cells and deliver them to lysosomes where most of the antimicrobial proteins reside.

The complement system consists of more than 20 different proteins circulating in blood that complement the function of antibodies and phagocytes in killing pathogens. Upon activation by microbial cell walls, the complement system leads to direct lysis of the microbial cells as well as to coating of the microbial cell (opsonization), marking it for destruction by phagocytes. The complement proteins are produced by the liver as acute-phase proteins, whose concentrations in plasma increase or decrease in response to inflammatory signals, most notably interleukins, which are secreted by macrophages in response to infection or injury.

Detection of viral infections leads to production of type I interferons by infected cells as well as by specialized hematopoietic cells called plasmacytoid dendritic cells. Type I interferons are essential for antiviral defense: they induce about 200 proteins with various antiviral activities, including inhibition of viral replication, of virion budding, and of viral protein synthesis.

When a multicellular parasite such as a helminth worm invades the host organism, the innate immune system deploys additional weapons. These weapons include mast cells and basophils, which produce histamine to promote blood flow to tissues and also have a prominent role in allergic disease; eosinophils, which fight parasitic infections by producing ROS, enzymes, and cytokines; and mucosal epithelium, which produces mucus to promote expulsion of parasites.

The innate immune system relies on pattern recognition receptors, including the ancient, widely shared Toll-like receptors, to sense microbial infections. These receptors detect conserved structures that are unique to microorganisms. Their activation induces an immediate inflammatory and antimicrobial response using the weapons mentioned above. Cytokines are proteins made by one set of cells that affect the behavior of other sets of cells. Major inflammatory cytokines, including interleukin-1 and interleukin-6 (IL-1 and IL-6) and tumor necrosis factor α (TNF-α) coordinate a complex inflammatory response involving liver, bone marrow, the hypothalamus, fat, and muscle. In addition, innate immune recognition of infection leads to activation of dendritic cells, which are sentinels posted in most tissues to detect pathogens, digest them, and present their antigens to T-lymphocytes so as to initiate the adaptive immune response (Figure 4.12).

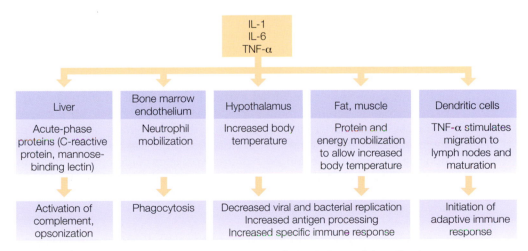

Figure 4.12   The innate immune system triggers the activation of the adaptive immune system through the production by phagocytes of the cytokines IL-1, IL-6, and TNF-α. (After Murphy et al. 2008.)

In summary, the innate immune system provides the first line of defense against infection. It is engaged immediately upon infection and some components of innate immunity are continuously active to prevent pathogen entry. In addition, the innate immune system instructs the adaptive immune system about the type of defense to produce for each class of pathogen: viruses, bacteria, fungi, or helminths.

The adaptive immune system

Adaptive immune systems are found only in vertebrates, having evolved independently at least twice, once in jawless fish (lampreys and hagfish) and once in jawed vertebrates. The two systems share some design features but clearly evolved independently, a pattern that underlines the functional advantages of the shared features.

In jawed vertebrates, the adaptive immune system appeared abruptly when a transposable genetic element carrying two genes, *RAG1* and *RAG2*, was incorporated into a genetic locus encoding immunoglobulin family cell-surface receptors. These genes carried the machinery that enabled the key feature of the adaptive immune system, the somatic recombination that now generates antibody and T-cell receptor diversity. This feature endows the system with the flexibility to recognize a huge range of pathogen molecules, some never previously encountered, because its output is a highly diverse repertoire of antigen receptors—immunoglobulins and T-cell receptors—whose range is so broad that they can detect any antigen, be it self, microbial nonself, or nonmicrobial nonself (e.g., food antigens). A key feature of the adaptive immune system is that the specificities of the antigen receptors are not genetically predetermined; instead, each receptor is assembled from gene fragments by a process of somatic recombination mediated by *RAG* genes.

MHC class I and class II molecules also first evolved in cartilaginous fish. Both types of MHC molecules display (present) fragments of proteins (antigenic peptides) on cell surfaces to be recognized by the T-cell receptors. The MHC class I molecules are found in all nucleated cells of the body. They present antigens derived from proteins synthesized within the presenting cell (e.g., viral proteins). MHC class II molecules are expressed on specialized cells, including dendritic cells, macrophages, and B cells, that present antigenic peptides derived from extracellular sources of proteins (e.g., phagocytosed bacteria). In most species, including humans, MHC class I and class II genes are highly polymorphic; they have many allelic forms at the population level. At the level of individual organism, these genes function in a codominant fashion, each of the two alleles conferring the ability to present different sets of antigenic peptides. In outbred populations, most individuals are heterozygous for MHC genes, which provides the obvious advantage of increased coverage of possible antigenic peptides derived from diverse pathogens.

Having briefly sketched the origin of the vertebrate immune system, which occurred about 500 mya, we now describe the immune organs, tissues, and cell populations. The bone marrow is a factory continually

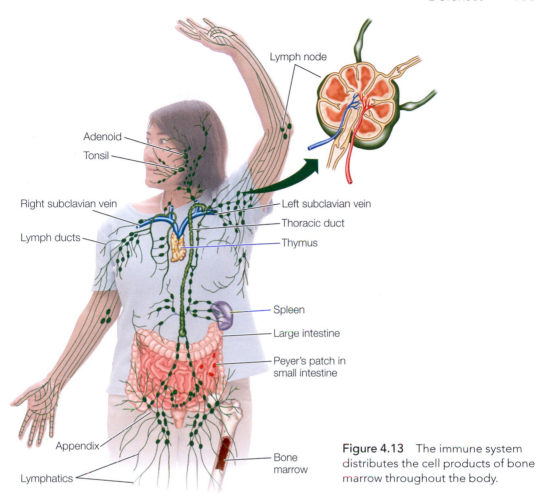

Lymph node

Adenoid

Tonsil

Right subclavian vein

Lymph ducts

Left subclavian vein

Thoracic duct

Thymus

Spleen

Large intestine

Peyer's patch in small intestine

Appendix

Bone marrow

Lymphatics

Figure 4.13   The immune system distributes the cell products of bone marrow throughout the body.

producing a variety of lymphocytes and leukocytes, as well as erythrocytes and platelets, from hematopoietic stem cells. Whereas *B* cells differentiate within *bone* marrow (in mammals), *T*-cell progenitors migrate to the thymus, where they undergo terminal differentiation. A key aspect of differentiation of both T and B lymphocytes is the screening of their antigen receptors for self-reactivity: lymphocytes that express receptors reactive to self-antigens are eliminated by apoptosis. Once fully differentiated, T and B cells migrate through the blood to various lymphoid organs, including lymph nodes, the spleen, the tonsils, the appendix, and Peyer's patches. These structures are parts of the lymphatic system, which drains tissues to remove excess interstitial fluids along with waste products and any foreign materials, including pathogens, should they occur in the tissues (Figure 4.13). The lymph is filtered in the lymph nodes before returning to the blood through the lymphatic ducts. T and B lymphocytes circulate via blood through multiple lymphoid organs until they encounter an antigen that can trigger their antigen receptors.

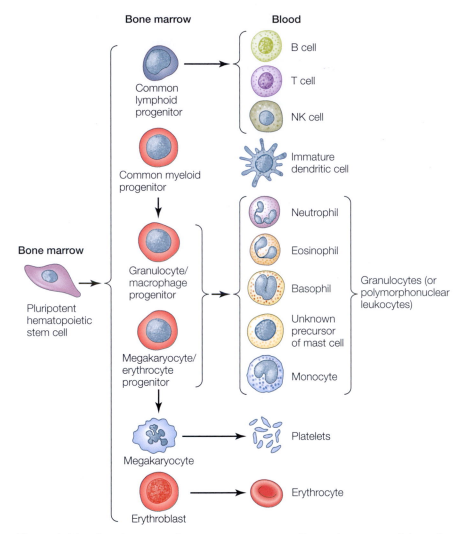

Figure 4.14 The pluripotent hematopoietic stem cells produce most of the cells found in blood and lymph, including all the types of immune cells. (After Murphy et al. 2008.)

When immune cells leave the bone marrow, they take one of two paths (Figure 4.14). The lymphoid progenitors of T cells move to the thymus to mature, as described above, while the lymphoid progenitors of B cells mature in place before moving to the lymphatic system. That is one path. Other cells, the myeloid progenitors, migrate to tissues throughout the body, where they differentiate into dendritic cells, blood monocytes, granulocytes, and macrophages.

When immature dendritic cells encounter a pathogen, they phagocytose it, process its antigens, display its fragments on MHC molecules on their plasma membrane, and migrate to lymph nodes, where they activate naive

T cells. As they migrate, they lose their ability to engulf pathogens and develop increased ability to stimulate T cells (Kareva and Hahnfeldt 2013). Once activated by dendritic cells in the lymph nodes, a T cell proliferates to establish a clone of cells with the same antigen receptor and differentiates into either cytotoxic "killer" T cells or "helper" T cells that activate B cells to produce antibodies and macrophages to make them microbicidal. In addition, helper T cells produce various cytokines that help orchestrate local immune responses through activation of eosinophils, basophils, mast cells, epithalia, and stromal cells.

Clonal selection

Clonal selection theory provides a conceptual framework of the functioning of the adaptive immune system. The elements of clonal selection are as follows. Each lymphocyte (T or B cell) bears a single type of antigen receptor with a unique specificity. The binding of a foreign antigen to a lymphocyte receptor activates the lymphocyte, which multiplies to form a clone of effector cells with identical antigen receptor specificities. Lymphocytes bearing receptors with high affinity for self-antigens are deleted early in development and are not found in the pool of mature lymphocytes (Figure 4.15A).

The diversity of lymphocyte antigen receptors in mammals is generated by somatic gene rearrangements, gene conversion, nontemplated nucleotide additions at the ends of gene segments, and random pairing of two polypeptides that form antigen receptors (Figure 4.15B). These processes, acting essentially at random, produce a huge array of diverse receptors, each expressed on a different lymphocyte clone.

An antibody (see Figure 4.5)—a protein that binds specifically to a particular substance, its antigen—is used to defend the host mainly in three ways. First, antibodies bind to viruses and bacterial toxins, neutralizing them by preventing viral entry into cells or by blocking the activity of toxins. Second, they coat—opsonize—the surface of pathogens so that they can be eaten and killed by macrophages. Third, they activate complement proteins that also coat extracellular pathogens for consumption by macrophages and, in some cases, kill pathogens directly.

Intracellular pathogens, which can potentially hide from the immune system inside cells, require special treatment. Cells infected by viruses use MHC class I molecules to present fragments of the viruses on their surfaces, activating cytotoxic T cells that kill them to prevent viral replication. Macrophages that have ingested leprosy pathogen use MHC class II molecules to transfer fragments of the bacteria from their phagosomes to their cell surface. In this activated state, they recruit helper T cells that detect the antigen on their surface, presenting it to B cells that are then recruited through clonal selection to produce specific antibodies in large quantities.

Immunoglobulin receptors are presented on the surface of B cells, each type of B cell only producing one specific antibody, which is a secreted version of the immunoglobulin receptor. When the antibody binds with a

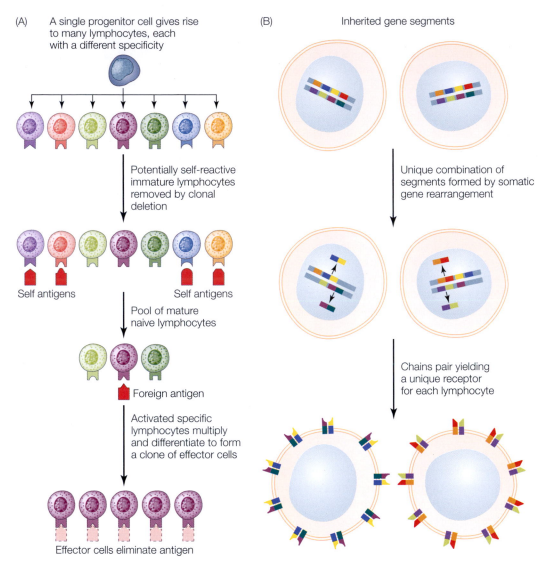

(A) A single progenitor cell gives rise to many lymphocytes, each with a different specificity

Potentially self-reactive immature lymphocytes removed by clonal deletion

Self antigens   Self antigens

Pool of mature naive lymphocytes

Foreign antigen

Activated specific lymphocytes multiply and differentiate to form a clone of effector cells

Effector cells eliminate antigen

(B) Inherited gene segments

Unique combination of segments formed by somatic gene rearrangement

Chains pair yielding a unique receptor for each lymphocyte

Figure 4.15 (A) The clonal selection hypothesis explains the flexibility, power, and precision of the adaptive immune system. (B) The diversity of MHC receptors is generated by somatic gene rearrangements. (After Murphy et al. 2008.)

specific antigen, it is endocytosed by the B cell and degraded into peptide fragments. These fragments bind to MHC class II molecules and are transported to the cell surface, where they are presented to T cells. A special type of helper T cells present in lymphoid organs then activates B cells to become antibody-producing cells after clonal expansion. A key consequence of the clonal design of the adaptive immune system is the generation of clones of long-lived memory lymphocytes that can become rapidly activated upon repeated encounters with the same pathogens. In most cases, these memory cells provide efficient protection from subsequent infections.

Immune defense and its costs

The immune system uses three main defensive strategies. It kills bacterial and fungal pathogens with phagocytes, antimicrobial peptides and proteins, and the complement system: it eats, poisons, or smothers them. It uses cytotoxic T cells to eliminate cells infected with viruses, intracellular bacteria, and protozoa. In addition, it expels multicellular parasites from the body using mucosal epithelia, smooth muscles, mast cells, and basophils.

Each of these defense strategies has an associated immunopathology and corresponding types of autoimmune and allergic diseases. Reactions mediated by immunoglobulin E and mast cells cause allergies. Reactions mediated by cytotoxic T cells, the complement system, and antibodies cause autoimmune anemia. Reactions mediated primarily by antibody-antigen complexes cause rheumatoid arthritis and lupus. In addition, T cells also play an important role in multiple sclerosis, psoriasis, and other autoimmune disorders.

■ SUMMARY

The innate immune system provides a rapid, generalized defense based on recognition of pathogen-associated molecular patterns shared by broad classes of pathogens. The adaptive immune system is activated by the innate immune system and relies on it to "know" what type of pathogen to which it is responding: viruses, bacteria, fungi, protozoa, or helminths. The key features of the adaptive immune system that make it so flexible and powerful are somatic generation of diverse antigen receptors and clonal selection of cells with the receptors that match pathogen antibodies. The adaptive immune system thus uses the principle of natural selection to shape its output. It also adds memory in the form of stored populations of B cells previously deployed against specific antigens. It is this immunological memory that makes vaccination effective.

Pathogen Evolution

Pathogens have their own agendas: they evolve rapidly in response both to changes in their individual hosts and to human interventions. The most important evolutionary responses of pathogens concern virulence, their role in the microbiota (which can switch from symbiotic to pathogenic), evasion and suppression of the immune system, resistance to antibiotics, and reactions to antipathogen interventions designed to minimize evolutionary responses. We now discuss these responses in sequence.

Virulence caused by pathogens

The detrimental effects of infections on hosts are caused both by the direct damage done by pathogens and by the hosts' responses to infection. Virulence is always an interaction between pathogen infection and host response. We use virulence in this chapter to refer to the portion of the damage that the infection causes the host that is due to properties intrinsic to the pathogen rather than to the host's reaction to the infection. The properties of the pathogen are determined by its evolutionary agenda, which is to maximize its reproductive success over its entire life cycle. That life cycle encompasses everything that happens from infection of a host to transmission into and infection of the next host, including the effects of host responses on pathogen survival and reproduction. Seen from the point of view of the pathogen, five key issues shape its virulence.

First, does its impact on the host affect its probability of transmission? If it kills the host too rapidly, it may not be transmitted at all; if it does not exploit the host efficiently, it may lose to competitors that make better use of host resources to produce transmittable progeny. This is the virulence-transmission trade-off.

Second, to what degree is it horizontally transmitted rather than vertically transmitted to offspring of this host? Once it is horizontally transmitted, its reproductive success no longer depends on the continued survival of its original host, whose survival becomes irrelevant, but if it is vertically transmitted, its reproductive success depends on the survival of its host until its host reproduces, which could happen several times.

Third, is it the only genetic strain or species of pathogen infecting the host, or are there other strains or species with which it must compete while attempting to reproduce and transmit? If it is alone in the host, it can evolve a level of virulence that maximizes transmission probability, but if it is in a coinfected host, it must scramble to use host resources before its competitors do, even if doing so reduces transmission probability below what it could achieve if it were the only pathogen present. This is the issue of single versus multiple infections.

Fourth, is its ability to infect this host a function of its ability to infect other hosts? We expect a jack-of-all-trades to be a master of none: specialists should outcompete generalists. Evidence supporting that view comes from the production of live attenuated vaccines through serial transfer, which causes the pathogen to increase in virulence on its new host while it loses virulence on its old host, rapidly becoming harmless enough on its old host to be used as a live attenuated vaccine. The ecosystem in which pathogens are evolving during serial transfer is a special one and may not represent natural conditions, in which host genotypes and phenotypes are both more variable.

Fifth, how much of this host has it seen in the past? Not all hosts and pathogens are locked in long-term arms races caused by repeated cycles of infection and transmission. Sometimes, pathogens jump into new hosts, producing emerging diseases, situations in which neither the host nor the pathogen has had much, if any, evolutionary experience of the other. Such spillover events are associated with unpredictable levels of virulence, ranging from harmless to catastrophic. In general, the precision of adjustment of pathogen to host depends on the frequency with which the two have encountered each other.

We now examine each of these five issues in more detail.

The virulence-transmission trade-off

The paradigmatic example of virulence evolving in a trade-off with transmission is the introduction of the myxoma virus to Australia to control rabbits. Rabbits are not native to Australia; they were introduced by European colonists. The rabbits escaped from their hutches, increased rapidly in an environment that lacked predators that could control them, and grazed the vegetation down to the ground over vast areas, causing devastating damage to farms and ranches. An international search for diseases that could control the rabbits located the myxoma virus in the Americas, where it causes tumors but not immediate death in the indigenous cottontail rabbits (genus *Sylvilagus*). The myxoma virus, a member of the poxvirus family and a relative of cowpox, monkeypox, and variola, is transmitted by anthropods, including mosquitoes, fleas, lice, ticks, and mites. In European rabbits (those introduced to Australia), it causes myxomatosis, an acute disease with an initial 100% mortality rate; this spillover event causes initial host mortality well above the optimum for the virus.

When the virus was introduced to Australia in 1950, the authorities wisely preserved control samples of virus and rabbits that did not then coevolve with each other. Later, they would use comparisons with the coevolved viruses and rabbits to determine to what degree the evolution of virulence was caused by changes in each (Fenner 1983). The initial introduction was a single, highly virulent strain that soon mutated into a diversity of competing, less virulent strains. These strains were classified into virulence grades ranging from I to V, with I being the most virulent, causing a case fatality rate greater than 99%, and V being the least virulent, causing a case fatality rate of less than 50%. Strain III with intermediate virulence rapidly outcompeted both the highly virulent initial strain I and the least virulent mutant strain V (Table 5.1). Although the extrinsic virulence of the virus did decline from its initial catastrophic level, the disease remained deadly after it stabilized at virulence level III, where it has a case fatality rate of 70%–95% and continues to function well in rabbit controls. Comparisons with unevolved strains demonstrated that the evolved level of virulence was the result of evolutionary changes both in the intrinsic virulence of the virus and in the ability of the rabbits to resist or tolerate infection.

If the rabbits die quickly, there is little opportunity for the arthropod vectors to transmit the disease. Less virulent strains then outcompete more virulent strains because of their superior transmission. Although they may do worse in a single host, they do better in the population as a whole. The process continues until a stable level of intrinsic virulence in the virus and of resistance/tolerance in the host population evolves (Figure 5.1). When that level of virulence is achieved, the disease is still very serious.

TABLE 5.1   The virulence of strains of myxoma virus in Australia from 1951 to 1981

Virulence grade:	I	II	III	IV	V	
Case fatality rate (%):	>99	95–99	70–95	50–70	<50	
Mean survival time (days):	<13	14–16	17–28	29–50	—	Number of samples
1950–1951	100					1
1952–1955	13.3	20.0	53.3	13.3	0	60
1955–1958	0.7	5.3	54.6	24.1	15.5	432
1959–1963	1.7	11.1	60.6	21.8	4.7	449
1964–1966	0.7	0.3	63.7	34.0	1.3	306
1967–1969	0	0	62.4	35.8	1.7	229
1970–1974	0.6	4.6	74.1	20.7	0	174
1975–1981	1.9	3.3	67.0	27.8	0	212

Source: After Fenner 1983, Table 4, p. 265.
Note: Data are expressed as percentage of samples recovered.

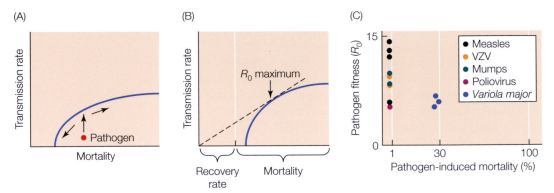

Figure 5.1 Trade-offs in virulence evolution. (A) A trade-off between transmission rate and host mortality; the pathogen will evolve upward to the boundary. (B) Parasite fitness (R_0) is then proportional to the sum of the recovery rate and the mortality rate; maximum fitness is achieved where the line through the origin is tangent to the trade-off curve, where the host suffers some mortality. (C) Data on R_0 and prevaccination mortality rates, however, show that some viral diseases result in negligible host mortality. VZV, varicella zoster virus. (After Bull and Lauring 2014.)

The virulence-transmission trade-off has been demonstrated with experimental evolution in beetles infected with microsporidians, where pathogen virulence decreases to an intermediate level and host genetic resistance increases (Berenos, Schmid-Hempel, and Wegner 2009). It has also been invoked to explain the decrease in virulence of syphilis in Europe following its introduction to Naples from the New World in 1495 (Harper et al. 2011). It is clear, however, that the simple virulence-transmission trade-off does not apply to all cases of host-pathogen interaction because at the evolutionary optimum, there should be some measurable level of host mortality (see Figure 5.1B). Instead, there are several infectious diseases, among them common and highly transmissible viral infections, in which the case fatality rate is very low (see Figure 5.1C; Bull and Lauring 2014). The solution to this paradox could be that it is the host recovery rate, not host mortality, that trades off with transmission, but we do not yet know if that is the case.

Vertical versus horizontal transmission

Strictly vertical transmission from parent to offspring selects for lower virulence, eventually reaching an evolutionary equilibrium in which the pathogen becomes an avirulent commensal. In contrast, strictly horizontal transmission selects for higher virulence until problems with transmission halt the increase. Selection on virulence in vector-borne pathogens depends on the impact of virulence on the efficiency of the vector. Many waterborne diseases are horizontally transmitted, infect large numbers of people, and cause many deaths (Table 5.2). That is why providing public supplies of clean water can have a greater effect on mortality than other public health measures.

TABLE 5.2   Examples of waterborne diseases with major impact

Disease	Annual cases	Annual deaths
Typhoid fever	12 million	130,000
Cholera	3 million	60,000
Amoebic dysentery	50 million	60,000
Rotavirus	> 2 million	450,000
Bacillary dysentery	165 million	520,000

Sources: Typhoid fever, Mogasale et al. 2014; cholera, amoebic dysentery, and bacillary dysentery, Lozano et al. 2012; rotavirus, Parashar et al. 2003 and Tate et al. 2012.

Single versus multiple infection

If the impact of the pathogens on the host is primarily expressed as mortality rather than decreases in growth and reproduction, multiple infections select for increased virulence because competition among strains for representation in the transmission event alters the balance of the virulence-transmission trade-off. In such a case, the pathogen has to dominate the competition to be transmitted; in doing so, however, it also damages the host, thereby increasing its intrinsic virulence. In contrast, if parasites have sublethal effects on the host, such as slower growth, and if these effects feed back onto all the strains of infecting parasites to reduce their rate of multiplication, multiple infections generally lead to lower virulence.

One versus many host species

The more host species that are regularly encountered, the less well adapted the pathogen will be to any one of them. That is because hosts differ in the problems they pose to pathogens, and a pathogen that evolves to be good at exploiting one host loses efficiency on others. This principle has been exploited—and demonstrated—in the production of attenuated live vaccines through serial passage.

Serial passage is an evolutionary technology that works because pathogens evolve rapidly to specialize on a new host when that is the only possibility given to them. The experimental host is chosen to be genetically uniform and is not allowed to coevolve: pathogens encounter the same naive host in every passage. Transmission costs are eliminated by the technology: pathogens are removed from one host while they are still in exponential growth phase and transmitted immediately into the next host, where they resume exponential growth. From the point of view of pathogens, they grow continuously and exponentially in genetically uniform, evolutionarily and immunologically naive hosts. Life is made as simple as possible for them as they are repeatedly transferred from one host to another. As a result, their virulence on the new, experimental host increases (Figure 5.2A), and their virulence on their original, natural host decreases. That is how Sabin, Hennessen, and Winsser (1954) produced live polio vaccine; after 50 passages in cultures of monkey kidney or testis cells, when it no longer caused any deaths when tested in monkeys not previously exposed, it was deemed safe for use in humans (Figure 5.2B).

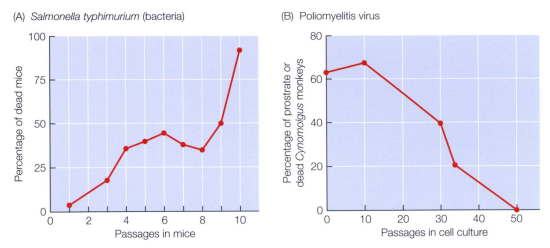

(A) *Salmonella typhimurium* (bacteria)

(B) Poliomyelitis virus

Figure 5.2   Serial passage increases virulence in the new, experimental host (A), whereas it decreases virulence in the original, natural host (B). (After Ebert 1998.)

Spillover virulence in facultative opportunists

Some pathogens—including those causing malaria and tuberculosis—live only or primarily in humans and are well adapted to us. Their level of virulence has been adjusted by coevolution to an intermediate but still damaging level at which they achieve successful transmission and can maintain themselves in our population. Others are facultative opportunists that spend most of their time in other hosts and whose level of virulence in humans has not been adjusted by a coevolutionary process to an intermediate level. When these pathogens infect humans, their virulence is unpredictable. The infections that cause little or no damage and are asymptomatic are doubtless numerous but go unnoticed. Those that are noticed cause highly virulent diseases, such as bubonic plague, Ebola, severe acute respiratory syndrome (SARS), the diseases caused by Hendra and Nipah viruses, Middle Eastern respiratory syndrome (MERS), and rabies. The natural hosts of plague are small rodents; those of Ebola, SARS, Hendra, Nipah, and MERS are bats and flying foxes; and rabies lives in a variety of mammals, including raccoons, foxes, and vampire bats. For these pathogens, which may be asymptomatic in their natural hosts, infecting humans is a dead end. They kill humans too rapidly to transmit effectively, and their virulence does not have time to equilibrate to a coevolutionarily stable level.

The effect of human interventions on pathogen virulence

Whenever humans intervene in the lives of pathogens, pathogens produce an evolutionary response. Any intervention that increases the efficiency of horizontal transmission (e.g., contamination of water supplies), that increases the frequency of multiple infections with impacts on mortality, or that selects for more virulent strains within hosts (e.g., imperfect vaccines) may cause the evolution of increased intrinsic virulence. In contrast, there is little we can do to alter the intrinsic virulence of pathogens like the Ebola and rabies viruses because their evolution is primarily determined by coevolutionary interactions with nonhuman hosts.

Imperfect vaccines and virulence

Because of their wide application in public health programs, vaccines are, with antibiotics, one of the two top human interventions in pathogen evolution. Childhood vaccines provide near-perfect protection from diseases like mumps, measles, rubella, and polio because they elicit sterilizing immunity. In contrast, imperfect vaccines, such as those being developed for malaria and currently in use for human papilloma virus, pose a threat because they are not completely sterilizing and allow some strains to transmit. Increased virulence is then expected to evolve for two reasons (Mackinnon, Gandon, and Read 2008; Huijben et al. 2013).

First, before the vaccines were applied, the average time that a patient survived was limited by the virulence of the circulating strains. Vaccinated patients will survive longer than unvaccinated ones, and that extension of their survival time will allow more virulent strains to survive and transmit. That is a simple, first-order effect of imperfect vaccines.

Second, at evolutionary equilibrium, the virulence of the pathogen is in part determined by the virulence-transmission trade-off. By providing partial protection that increases the survival of infected hosts, vaccination reduces the mortality cost of virulence in the vaccinated hosts, thus shifting selection to a new equilibrium at a higher intrinsic virulence in the vaccinated host population. This level of virulence is not expressed in the vaccinated hosts, which are partially protected, but if the evolved pathogen strains then infect naive, unvaccinated hosts, they will be more virulent than the strains that circulated before vaccines were used.

The expected increase in pathogen virulence when imperfect vaccines are used is not a reason not to vaccinate. Even an imperfect vaccine for a disease like malaria will save millions of lives. It is, however, useful to know about, and therefore be able to plan for, the predicted evolutionary consequences.

■ SUMMARY

The intrinsic virulence of pathogens—the portion of the damage that the infection causes the host that is due to properties intrinsic to the pathogen rather than to the host's reaction to the infection—is determined by the relative strengths of at least five different effects: the trade-off of virulence with transmission, the balance of vertical versus horizontal transmission, the frequencies of single versus multiple infections, the number of different types of hosts regularly infected, and the biology of the pathogen when humans are a novel host colonized by a spillover event. Human interventions that can affect virulence evolution in pathogen populations should be considered with care.

Managing the microbiome: Symbionts versus pathogens

The traditional view is that the function of the immune system is to defend the body against invasive pathogens. That certainly remains one of its primary functions, but now it is seen as having another, perhaps equally important, one: managing the microbiota.

We now know that humans are colonized by more than 1000 species of bacteria and by an unknown, certainly large number of viruses. These commensal microorganisms do not normally cause disease in a healthy host with an intact immune system, but some of them can do so in immunocompromised hosts or if they change their niche within the body. Within the normal microbiota, pathogens represent a small fraction, whereas many of the commensal microbes provide essential benefits to the host: they produce vitamins and help to digest complex polysaccharides, they detoxify some harmful chemicals, they compete with pathogens for intestinal niches, and they produce the molecular signals necessary for proper maturation of the immune system in the developing infant.

The immune system helps maintain a healthy symbiotic relationship with the commensal microorganisms that constitute our microbiota. We each consist of about 10 trillion mammalian cells, about 90 trillion bacterial cells, and an unknown, large number of viruses. That the immune system plays an important role in managing our relations with our symbionts is revealed by the infections that develop in people who are immunocompromised. It does so by controlling the growth and invasion of the commensal microbes capable of causing disease in the immunocompromised state: the opportunistic endogenous pathogens. Although the interaction interface occurs wherever mammalian cells encounter microbial cells, the action is most intense at the epithelial mucosae where we need to gain access to oxygen and nutrients, thus opening doors through which other things can enter as well. The epithelium is one such site.

The gut ecosystem

The host and its immune system encourage colonization by beneficial bacteria (Figure 5.3). The oligosaccharides that are the third most abundant component of breast milk feed Bifidobacteria that stimulate the developments of the infant's immune system. When *Bifidobacterium longum infantis* is grown on the oligosaccharides found in human breast milk, it adheres better to cultured human intestinal epithelial cells than when it is fed lactose. There it also induces greater expression of an anti-inflammatory cytokine (IL-10) and of junctional adhesion molecules than do control strains grown on lactose (Chichlowski et al. 2012).

In the gut and other colonized sites, mucosal secretory cells produce defensins—peptides that are active against bacteria, fungi, and viruses—and immunoglobulin A (IgA), 3–5 g of which are secreted into the gut lumen each day, forming a protective boundary layer. Several pathogens have evolved countermeasures; for example, *Neisseria gonorrheae*, *Steptococcus*

(A) Healthy gut environment

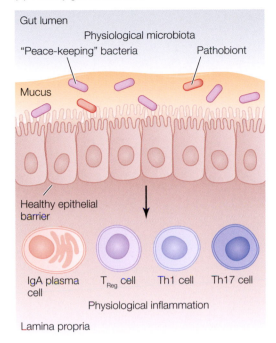

(B) Altered gut environment

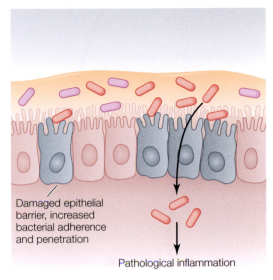

(C) Altered host immune system

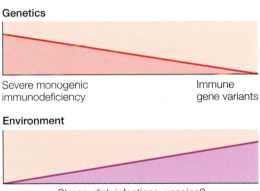

Figure 5.3 (A) In the healthy gut, microorganisms are maintained as commensals in the gut lumen by the action of the immune system. Some physiological inflammation is a normal part of how the immune system manages the gut microbiota. (B) If the defenses are breached, immune activation results in additional, pathological inflammation, and infection may follow if the invading bacteria are not rapidly controlled. (C) The host immune system can be altered both by genetic inheritance and by environmental factors, which combine to produce various mixtures of causation. IgA, immunoglobulin A; T_{Reg}, regulatory T cell; Th, helper T cell. (After Cerf-Bensussan and Gaboriau-Routhiau 2010.)

pneumonia, and *Haemophilus influenza* type B all produce a protease that destroys IgA. The advantages of defense are, as usual, accompanied by costs, for example, deposits of IgA can cause chronic kidney disease, and immune interactions with IgA are involved in celiac disease.

The skin ecosystem

The skin is not only a defensive barrier; it is also a complex ecosystem colonized by microbiota that are in dialogue with the host immune system (Figure 5.4). Our skin hosts about 1 million bacteria per square centimeter,

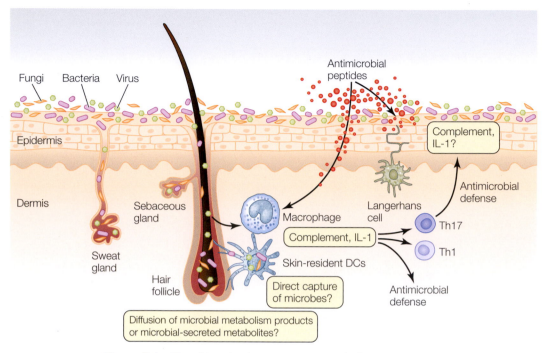

Figure 5.4 The skin microbiota converse with the immune system, eliciting the production of complement and interleukins. The boxes highlight potential communication mechanisms; question marks indicate those not yet confirmed. IL, interleukin; DC, dendritic cell; Th, helper T cell. (After Belkaid and Segre 2014.)

a total of about 10^8 bacteria per human constituting a diverse set of communities organized primarily by skin physiology (moist, dry, or sebaceous). These communities are colonized in two waves: first in infants and young children, and then at puberty, when lipophilic bacteria move in and displace the prior residents.

Both at steady state and during infection, the skin microbiota function as an endogenous adjuvant that supports and promotes immunity by enhancing the activation of lymphocytes. Healthy human skin contains about 20×10^9 lymphocytes and is one of the largest reservoirs of memory T cells in the body (Belkaid and Segre 2014). When the skin microbiota are disturbed, however, skin disorders—including psoriasis, atopic dermatitis, and acne—can result. Atopic dermatitis (AD) is twice as prevalent in industrialized countries as it is in developing countries, and more than half of the children that suffer from moderate to severe AD also develop hay fever or asthma. The mechanisms through which the skin microbiota either initiates or amplifies skin disorders are not yet well understood, but plausible working hypotheses have been suggested (Figure 5.5).

Extrinsic and intrinsic virulence

When a pathogen is virulent in one host but not in another, the virulence is referred to as extrinsic. Many pathogens are relatively harmless in the

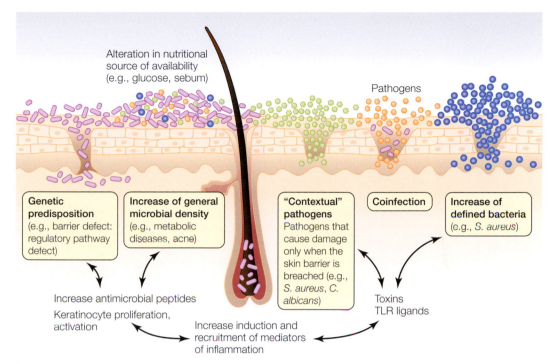

Figure 5.5   Five potential mechanisms with which the skin microbiota could initiate or amplify skin disorders; they may often act in combination. TLR, Toll-like receptor. (After Belkaid and Segre 2014.)

organisms in which they usually live but are highly virulent in accidental hosts into which they spill over. For example, the Ebola virus is thought to be harmless in its natural host, fruit bats, but is lethal in primates; similarly, the viruses causing SARS and MERS and the disease caused by Nipah virus are relatively harmless in their natural reservoir hosts, among which bats and flying foxes feature prominently, but they cause high rates of death when they emerge as human diseases.

In contrast, intrinsic virulence is defined by the presence of virulence genes that express toxins in the pathogen that make it more virulent than similar microbes that lack such genes. Such is the case with diphtheria, whose symptoms are caused by toxins produced by *Corynebacterium diphtheria*, and tetanus, caused by a toxin released by *Clostridium tetani*. Toxins are public goods in the sense that all bacterial cells benefit from them, whether they are producing them or not, which causes less evolutionary response of the bacteria to vaccination; we discuss it in more detail below.

In some cases, evolutionary changes in virulence can be examined by comparing closely related species pairs that differ in virulence. The influenza virus is more virulent than the rhinovirus that produces the common cold; the bacterium that produces anthrax, *Bacillus anthracis*, is more virulent than its relative *Listeria monocytogenes*; and the bacterium responsible for bubonic plague, *Yersinia pestis*, is more virulent than its close relative, *Y. enterocolitica*, from which it fairly recently evolved (Parkhill 2008).

Extrinsic and intrinsic virulence are both products of interactions between host and pathogen best measured and conceptualized in cases in which the coevolutionary trajectories of virulence and resistance can be followed over time. That was the case with myxomatosis discussed earlier in this chapter.

Tolerance, resistance, and susceptibility: The lessons of rinderpest

Pathogens can evolve to be more intrinsically virulent or less intrinsically virulent, and hosts can evolve to be more resistant or more tolerant. Both types of changes often happen at the same time in coevolution, and both can contribute to establishing a symbiotic relationship. When the intrinsic virulence of symbionts is concealed by the evolution of tolerance in long-colonized hosts, there are important consequences for naive hosts that lack that coevolutionary history. Such consequences played out dramatically in the introduction of rinderpest from Asia to Africa.

Rinderpest, a virus related to the measles and canine distemper viruses, causes disease in cattle, buffalo, antelope, giraffes, warthogs, and bushpigs. Endemic to the Eurasian steppes, rinderpest was repeatedly introduced to Europe as a fellow traveler in human invasions, causing cattle plagues. It entered sub-Saharan Africa in cattle sent to relieve General Charles George Gordon in Khartoum in 1884 or introduced to Italian Somaliland in 1889. When it entered populations of wild ungulates around 1890, it spread rapidly through organisms with no evolved resistance. Between 1890 and 1899, it expanded into eastern, central, and southern Africa, eliminating most domestic cattle and wild buffalo and decimating the populations of many other species of wild ungulates. Subsequent outbreaks of rinderpest occurred in 1917–1918, 1923, and 1938–1941. One species of antelope went extinct, and the distributions of the other species were altered for a century.

There were consequences for humans. The pastoral and nomadic peoples lost their food sources and, under the stress of starvation, became susceptible to infectious diseases, suffering outbreaks of endemic smallpox. Over much of the infected area, tsetse flies—which transmit sleeping sickness, feed on ungulates, and require brush thickets as refuge—disappeared. As game disappeared, starving lions switched to eating people. In Uganda in the 1920s, one lion killed 84 people. Where lions were an increased threat, farmers abandoned their fields, in which thickets of brush then grew. The wild ungulates developed immunity to rinderpest and moved back into the abandoned farming areas, where they became hosts for the tsetse flies that could now survive in the new thickets of brush. Because the flies transmitted sleeping sickness, the human population withdrew further and remained absent even after the lions switched back to feeding on ungulates. Some of those areas then became national parks.

Although rinderpest was eliminated by 2001 through a vaccination campaign, it nevertheless changed the ecology of half a continent for at least a century. The consequences for humans were drastic, indirect, and the result of complex causal chains that could be understood after the fact but not predicted in advance. Rinderpest did to African ungulates what measles and smallpox did to Native Americans when Europeans brought those diseases with them to the New World.

■ SUMMARY

The immune system both defends against pathogens and manages relations with the microbiome. Long coevolutionary interactions between humans and their microbiome have produced physiological mechanisms that foster symbiotic relations. They include the production of oligosaccharides in breast milk that stimulate bifidobacteria to boost healthy immune responses in infant guts and the induction of the development of gut-associated lymphoid tissue (GALT, see Chapter 2) by signals emitted by gut bacteria.

Virulence is the product of host-pathogen coevolutionary and physiological interactions that have components contributed by both pathogen and host. We can see the consequences of coevolutionary adjustment of host-pathogen relations most clearly in cases in which it had not yet occurred, as when rinderpest invaded Africa.

Evasion and suppression of the immune system

The vertebrate immune system, a powerful and effective weapon that kills pathogens efficiently, causes strong selection on pathogen variants favoring strains that can evade or suppress it. The long coevolutionary history of interactions between hosts and pathogens has added many layers of complexity to responses on both sides, with virtually every response on one side provoking a counterresponse on the other. The result is that some pathogen has managed to exploit virtually every chink in the immune system.

Some pathogens confuse the immune system by varying their surface properties. The manner in which they do so depends on the group to which they belong. Influenza viruses undergo antigenic shift (recombination of entire DNA segments coding for surface proteins) and antigenic drift (point mutations within genes coding for surface proteins). Antigenic shift is a major change signaling the emergence of a new strain from animal reservoirs; antigenic drift is a relatively minor change in a strain already circulating in humans. Both shift and drift produce some adaptive mutations that increase in frequency under immune challenge (Luksza and Lässig 2014). Selection maintains serotype diversity in bacterial populations through selection/mutation balance and negative frequency dependence. In addition, some bacteria generate variation in surface properties through mechanisms that vary the number of simple sequence repeats, thereby producing frame shifts affecting gene expression. The eukaryotic malaria parasites, *Plasmodium*, and sleeping sickness parasites, *Trypanosoma*, both have sophisticated mechanisms that generate antigenic switching by varying expression among members of large gene families.

Other pathogens hide inside the genome or inside cells. Retroviruses can insert themselves into the host genome and hide there in a latent state.

Intracellular parasites hide within cells from immune weapons circulating in lymph and blood. Intracellular parasites include stages in the complex life cycles of the eukaryotes *Plasmodium*, *Trypanosoma*, and *Toxoplasma* as well as the mycobacteria that cause tuberculosis and leprosy and the bacterium *Listeria* that causes a relatively rare but unusually virulent type of food poisoning.

Pathogens can also disrupt immune function in several ways. They can block intercellular signaling or intracellular defenses, slow the recruitment of immune cells, or kill immune cells directly.

Immune suppression by viruses

Some viruses inhibit humoral immunity by disrupting receptors of signals of viral presence; they include herpes simplex, which causes cold sores and genital herpes; *Cytomegalovirus*, also in the herpes family, which infects more than 90% of humans worldwide and can cause serious infections in the newborn and the immunocompromised; and *Vaccinia*, a member of the poxvirus family related to cowpox and smallpox. Viruses in the herpes family can also block antigen processing by inhibiting gene expression and peptide transport, thus preventing the presentation of major histocompatability (MHC) class I molecules that act as signals of viral presence on the cell surface, and they can remain latent in cells for long periods.

TABLE 5.3 Composition, location, gene designation, and function of simple sequence repeat contingency loci in several genera of bacterial pathogens

Genus	Nucleotide composition	Location of repeat unit (no. of repeats)	Gene	Function of repeat unit
Yersinia pestis	5′-A-3′	Within ORF (9)	*yadA*	Membrane protein
Campylobacter jejuni	5′-C-3′	Within ORF (8)	*wafN*	Lipo-oligosaccharide synthesis
Bordetella pertussis	5′-C-3′	Promoter (15)	*fimB*	Adhesin
Helicobacter pylori	5′-CT-3′	Within ORF (8)	HP722	Membrane protein
Haemophilus influenzae	5′-TA-3′	Promoter (9)	*fim A, B*	Pilus
Escherichia coli	5′-TCT-3′	Within ORF (5)	*ahpC*	Stress response
Mycoplasma gallinarium	5′-GAA-3′	Promoter (12)	pMGA	Adhesin
Moraxella catarrhalis	5′-CAAC-3′	Within ORF	*tbp*	Unknown
Neisseria gonorrhoeae	5′-CTCTT-3′	Within ORF (7)	*opa*	Adhesion/invasion
H. influenzae	5′-GACGA-3′	Within ORF (4)	*hsd*	Restriction-modification
H. influenzae	5′-ATCTTC-3′	Promoter (16)	*hmw*	Adhesin

Source: After Moxon, Bayliss, and Hood 2006, Table 1, p. 309.
Note: ORF, open reading frame.

Some viruses inhibit inflammatory responses by disrupting cytokine signaling and cell adhesion; they include *Vaccinia*, *Myxoma*, and the Epstein-Barr virus. Some of the effects of the Epstein-Barr virus—which causes infectious mononucleosis and is associated with several cancers (lymphomas and nasopharyngeal carcinoma) and increased risk of several autoimmune diseases, including rheumatoid arthritis and multiple sclerosis—are mediated by the production of interleukin mimics that immunosuppress the host.

In the Epstein-Barr virus, those effects act between cells. The paramyxoviruses that cause mumps and measles, however, act within cells to hide their presence by disrupting sensors of viral RNA.

Variation of surface properties by bacteria

Many bacteria, including those causing bubonic plague, whooping cough, and stomach ulcers, vary their surface properties with a genetic mechanism based on simple sequence repeats (SSRs), either within the promoter or in the open reading frame (ORF), that are contingently activated by signals of immune attack (Table 5.3). The mechanism uses mispairing of SSRs to cause frame shifts, resulting in amino acid substitutions (Figure 5.6).

The four types of cell surfaces that result in *H. influenzae* illustrate the phenotypic consequences (Figure 5.7): a sialic acid, a galactose, or a phosphorylchlorine can be added or dropped, yielding four cell types. Each elicits a different set of immune molecules produced by a different host cell population.

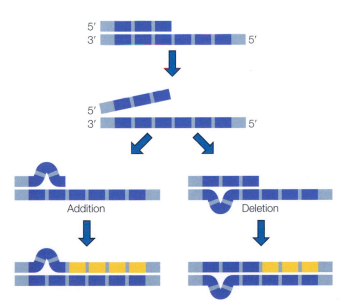

Figure 5.6 The mispairing of simple sequence repeats causes frame shifts, resulting in amino acid substitutions in bacterial surface proteins. (After Moxon, Bayless, and Hood 2006.)

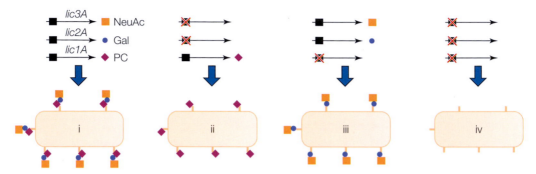

Figure 5.7   Variation in the cell surface of *H. influenzae*. NeuAc, a sialic acid; Gal, galactose; PC, phosphorylchlorine. *lic3A*, the gene mediating attachment of NeuAc; *lic2A*, the gene mediating attachment of Gal; *lic1A*, the gene mediating attachment of PC. (After Moxon, Bayless, and Hood 2006.)

Bacteria that hide inside cells

Infectious bacteria are often ingested, killed, and processed by macrophages, but some manage to convert their predators into hiding places. *L. monocytogenes*, which causes a disease whose symptoms are fever, muscle aches, and diarrhea, can move from macrophage to macrophage without ever emerging into the humoral environment. *Mycobacterium leprae*, which causes leprosy, spreads through the body by converting nerve cells to stem cells that then invade muscle, a particularly sophisticated hijacking of the mechanisms of gene expression that determine cell type.

Toxoplasma, *a single-celled eukaryotic pathogen, hides in cysts*

As many as a third of humans may have been exposed to and infected by *Toxoplasma gondii* through their contact with cats, the definitive host within which *Toxoplasma* reproduces sexually. *Toxoplasma*, which like *Plasmodium* and *Babesia* is a member of the Apicomplexa and shares with them a complex life cycle with a succession of stages, also infects many other warm-blooded animals. Infected cats excrete its eggs in their feces; transmission also occurs when a cat ingests an infected prey. Once in a new host, the eggs hatch and develop into a series of stages, the last of which forms a cyst containing hundreds of cells, usually in brain, liver, or muscle. Inside that cyst, the cells are protected from immune attack. *Toxoplasma* is also well known for its ability to manipulate the behavior of prey. Infected rodents are attracted to, rather than repelled by, the scent of cat urine, increasing the probability of being eaten and thus transmitting *Toxoplasma* cysts that burst in the cat's intestine, releasing an infective stage that burrows into the bloodstream.

Plasmodium *confronts and solves problems in two hosts*

Like other vector-borne pathogens, the malaria pathogen, *Plasmodium*, must deal with the immune system of its definitive host species (its mosquito

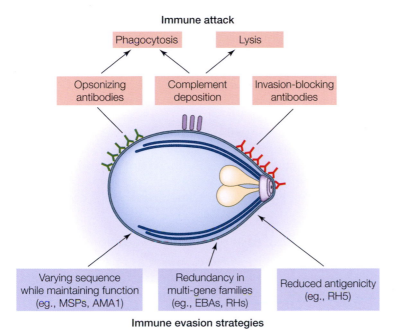

Figure 5.8   *Plasmodium* merozoites, the only stage of the malaria pathogen that is exposed to immune attack in the blood, evade attack with variable surface properties generated by both allelic polymorphism and variable expression from multigene families. MSPs, merozoite surface proteins; AMA1, a protein used in entering the red blood cell; EBAs, erythrocyte-binding antigens; RHs, erythrocyte-binding protein homologs. (After Wright and Rayner 2014.)

vector, in which it reproduces sexually) as well as that of its intermediate host species (vertebrates, in which it produces a series of asexual stages that infect different habitats). In mosquitoes, where it is attacked by melanization and phagocytosis, it coats itself to mimic the surface of mosquito cells and suppresses the mosquito immune response. In vertebrates, the stages exposed to humoral immunity, which attacks with complement and antibodies, vary their surface properties with some antigens that are highly polymorphic and others that are coded by multigene families with variable expression profiles (Figure 5.8).

One stage hides inside red blood cells. Red blood cells are a particularly good place to hide because they do not carry MHC class I proteins, cannot present signals of infection on their surface, and are therefore not subjected to attack by cytotoxic T cells. Infected red blood cells, however, are more rigid than uninfected cells, and that rigidity is recognized when they pass through the spleen, which filters them out of the blood. *Plasmodium* counters filtration in the spleen by inserting proteins into the membrane of the red blood cells in which it is hiding. Doing so helps them bind to capillaries, thus delaying their passage through the spleen, and as a side effect causes the blockage of blood vessels in the brain that cause cerebral malaria.

The protein inserted into the membrane of red blood cells to delay passage through the spleen, PfEMP1, elicits an antibody response that eliminates red blood cells containing *Plasmodium*, but *Plasmodium* counters that response by producing antigenic variation in PfEMP1, which is coded by a family of about 60 genes scattered over its genome. Only one gene is expressed at a time, and changes in which gene is expressed occur at a rate of about 2% per infected red blood cell. That is enough to keep the immune response out of phase with the parasite; the generals continue to fight the last war, which is steadily changing into a new war.

These measures and countermeasures document a long history of intense coevolutionary interactions between *Plasmodium* and the vertebrate immune system. This history has contained strong selection on both host and pathogen.

Trypanosomes, the classical case of antigenic variation

Trypanosomes are single-celled eukaryotes, not at all related to Apicomplexans, that cause three major human diseases: sleeping sickness, Chagas disease, and Leishmaniasis. They live extracellularly in host blood, where they are targeted by antibodies. Their surface is covered with a dense coat of a single glycoprotein, variable surface glycoprotein (VSG), that is frequently modified by changes in gene expression that recoat the surface with a new version of VSG. The trypanosome genome contains more than 1500 VSG genes, most in silent arrays. Most of them are pseudogenes that are activated by unknown recombination mechanisms that place them into one of the 15 telomeric VSG expression sites, only one of which expresses a protein at any one time. As with *Plasmodium*, this continual variation in surface proteins creates a situation in which the immune system is always gearing up to fight an enemy that has already mostly disappeared.

Immune suppression

Various pathogens use different methods to directly suppress immune function. *Mycobacterium tuberculosis*, which is an intracellular parasite, blocks the intracellular fusion of phagosomes with lysosomes, thus keeping itself from coming into contact with the enzymes that would digest it. The stage of *Plasmodium* that lives in red blood cells is able to interfere with and slow down the recruitment of T cells. *Leishmania* lives inside neutrophils and dendritic cells, both part of the cell populations of the immune system, and blocks their maturation. It also expresses an extracellular protein that disrupts immune signaling.

Many pathogens, ranging from viruses to worms, produce enzymes that disrupt immune function. For example, *Toxoplasma* phosphorylates a host resistance protein, and hookworms suppress the secretion of intestinal proteases and stimulate the production of immune cells that reduce inflammation. Particularly effective disrupters of immune function are cysteine kinases, a defensive solution on which viruses, bacteria, trypanosomes, and apicomplexans have converged. These enzymes cleave immunoglobulin G in blood and lymph, modulate concentrations of interleukins, and help control populations of immune cells.

■ SUMMARY

Most pathogens have evolved mechanisms to evade or suppress the immune system. Some vary their surface properties, some hide inside cells or inside the genome, some manipulate signals between cells in the immune system, some cleave antibodies directly, and some kill immune cells. Their abilities to carry out these functions are compromised by trade-offs with other functions necessary to pathogens, rendering imperfect their evasion and suppression of host immune systems. That imperfection helps constrain the evolution of disease virulence and allows many hosts to survive.

The rapid evolution of antibiotic resistance

Pathogens rapidly and repeatedly evolve resistance to antibiotics, particularly in the emergency rooms and intensive care units of hospitals where physicians try to keep patients in readiness for potential surgery. Often, patients enter a hospital without a bacterial infection, are infected by a bacterium that is resistant to most antibiotics, and then die from infection with a strain of bacterium that has evolved to become untreatable. Infections acquired in hospitals are called *nosocomial*. In 2004, more than 90,000 people in the United States died of nosocomial infections, more than died from traffic accidents, breast cancer, or AIDS. In that year, the cost of treating resistant infections exceeded $80 billion in the United States and probably approached $1 trillion worldwide. The supply of new antibiotics is not keeping pace with the rate at which pathogens are evolving resistance. If the pathogens that normally inhabit hospitals evolve resistance to all available antibiotics, surgery will become much more difficult and risky, and patients will die of bacterial infections for many other reasons as well.

Antibiotic resistance in clinically important bacteria has rapidly evolved every time a new antibiotic has been introduced (Table 5.4). For example, if a new antibiotic is introduced in northern England, resistance can evolve there within 6 months and turn up in Hong Kong within about 2 years.

This process creates a coevolutionary arms race between bacteria and the pharmaceutical industry. One particularly dangerous bacterium is multiply-resistant *Staphylococcus aureus*, a so-called superbug. Penicillin became available in 1943; staph evolved resistance to penicillin by 1947. In the 1960s, methicillin was introduced; by the 1990s, 35% of staph strains were also resistant to methicillin. In the 1990s, vancomycin was introduced; by 1996, staph had evolved resistance to vancomycin. In 2000, linezolid was introduced; by 2002, staph had evolved resistance to linezolid.

Costs have increased dramatically with resistance. In the United States, the cost of treating penicillin-resistant staph infections acquired in hospitals has been about $2 billion to $7 billion per year, the cost of treating

TABLE 5.4 The rapid evolution of resistance in clinically important bacteria

Antibiotic	Year introduced	Year resistance observed
Penicillin	1943	1945
Chloramphenicol	1949	1950
Erythromycin	1952	1956
Methicillin	1960	1961
Cephalothin (1st-generation cephalosporin)	1964	1966
Vancomycin	1958[a]	1986
2nd- and 3rd-generation cephalosporins	1979, 1981	1987
Carbapenems	1985	1987
Linezolid	2000	2002

Source: After Bergstrom and Feldgarden 2008, Table 10.1, p. 126.
[a]Although vancomycin was introduced in 1958, it was not widely used until the early 1980s.

methicillin-resistant staph infections acquired in hospitals has been about $8 billion per year, and the cost of treating staph infections in nonhospital (community) settings has been about $14 billion to $21 billion per year. The total cost of treating antibiotic resistance in staph has been $24 billion to $36 billion per year. Another example drives home the point: the cost of treating one case of nonresistant tuberculosis is about $25,000, whereas the cost of treating multiply-drug-resistant tuberculosis is about $250,000, or ten times as much. The current cost of treating all antibiotic-resistant bacterial infections worldwide is on the order of $1 trillion a year.

It helps to put that frightening picture in historical perspective (Figure 5.9). In the United States, the mortality rate from infectious disease fell steadily from 1900 to 1999 (with a dramatic spike caused by the 1918 influenza pandemic). Before the introduction of sulfonamides around 1930,

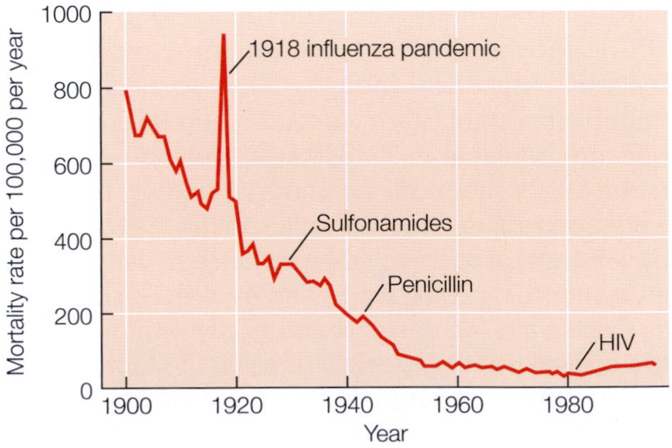

Figure 5.9 Mortality from infectious disease declined dramatically in the United States after 1900. Only part of that decline was caused by antibiotic therapy. (After Armstrong, Conn, and Pinner 1999.)

clean water, hygiene, and other improvements in public health and medical care had cut mortality rates from infectious disease in half. Subsequent reductions in mortality were due both to widespread use of childhood vaccines as well as use of antibiotics. Thus, if antibiotic therapy were to fail entirely, we would experience a resurgence in mortality rates, but they would probably not go back to the level experienced in 1930 because vaccination would still work and we would still have clean water supplies.

What are antibiotics, and how do they work?

Antibiotics are a heterogeneous class of molecules that interfere with the growth, survival, and reproduction of bacteria. Many antibiotics occur naturally and are isolated from bacteria and fungi. That is important because it points to a long coevolutionary history in which one species has produced molecules to mediate its competition with another species and the second species has reacted by evolving resistance. That often happened in the soil ecosystem, which has both high diversity and high density of bacteria and fungi. Over millions of years, it has produced an immense library of resistance genes capable of handling a broad diversity of antibiotic molecules. Thus, many, but not all, types of resistance evolved long before humans invented pharmaceutical companies. Naturally occurring antibiotics are often modified to increase their effectiveness, and some antibiotics are entirely synthetic, with no natural counterpart.

Antibiotics work in a variety of ways; most of them mimic a critical molecule, bind irreversibly to an active site, or compete with a naturally occurring molecule for binding, passage, or transport. The objective is to slow the growth of a target organism or kill it with minimal toxicity to the individual host. That balancing act is usually accomplished by directing the antibiotic at some phylogenetically unique feature of the target, some feature that only viruses, or only bacteria, possess and that patients do not have: for example, the bacterial cell wall, or the bacterial ribosome, which has some unique features that eukaryotic ribosomes do not share. Helminth worms, however, are multicellular eukaryotes whose biochemistry and cell structure is very similar to that of humans. That is why antihelminthic drugs are often so toxic to humans and require a careful therapeutic regime that manages to kill the worms before killing the patient.

How do resistance genes get into patients?

Many bacteria have a circular chromosome anchored to their cell wall and smaller, circular pieces of gene-bearing DNA called plasmids in their cytoplasm. A few have a linear chromosome; some do not have plasmids. Bacteria reproduce by asexual division, but they also engage in horizontal genetic exchange. When resistance has previously evolved in natural ecosystem, those resistance genes can move into human bacterial pathogens through any of three mechanisms (Figure 5.10).

The first is transformation, the uptake of naked DNA that is both eaten for its nutritional content and occasionally used as a source of genetic material. The subtitle of an article on transformation captured its essence: "Is sex

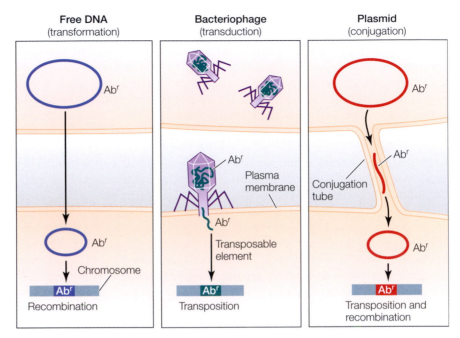

Figure 5.10 Bacteria can acquire resistance genes through any of three mechanisms: transformation, transduction, and conjugation. Ab^r, antibiotic resistance determinants. (After Alekshun and Levy 2007.)

with dead cells ever better than no sex at all?" (Redfield 1988). The second is transduction, which is mediated by viruses (bacteriophage) that bring along a piece of the genome of their last bacterial host when they infect a new one; they have evolved the enzymatic machinery to insert DNA into bacterial chromosomes. The third is conjugation, which is usually initiated by genes on plasmids; a bridge is built between two bacterial cells, allowing plasmids and sometimes chromosomes to cross over and recombine. The first two mechanisms, transformation and transduction, can move genes among phylogenetically distant species. The third, conjugation, moves genes within species.

All three methods of introducing foreign DNA into cells can result in incorporation of that information into the bacterial chromosome through homologous recombination. Transduction, which brings with it some viral enzymatic machinery, can also use that mechanism to insert DNA into bacterial chromosomes. And, information on plasmids can also be exchanged with information on chromosomes through transposition. The important point is that genetic information on antibiotic resistance moves frequently among both closely and distantly related bacteria.

Examination of bacterial genomes has suggested a useful distinction between core and accessory genes. The core genes are mostly involved in housekeeping functions like metabolism and biosynthesis; they form a stable, asexually inherited core that defines a strong phylogenetic signal

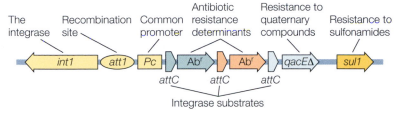

Figure 5.11   Transposable elements that confer antibiotic resistance. *int1* is the gene coding for the integrase, *att1* is the primary recombination site, *attC* are substrates of the integrase, and *Pc* is a common promoter that drives high-level expression of antibiotic resistance determinants (Ab[r]), including those against aminoglycosides, β-lactams, chloramphenicol, and trimethoprim. *qacEΔ* and *sul1* specify genes for resistance to quaternary ammonium compounds and sulfon-amides. (After Alekshun and Levy 2007.)

of relatedness. The accessory genes, the ones that are often horizontally exchanged, are involved in ecological interactions that can change in the short term: antibiotic resistance, pathogenic virulence, and nutrient uptake. When several such genes are moved in a block, they form what is called a pathogenicity island or resistance cassette. Both terms describe genes that have similar, often complementary functions and are located next to each other, forming units that integrate stably into the bacterial genome. The horizontal movement of such units delivers multiple genes for drug resistance in a single exchange (Figure 5.11).

Bacterial genetics have implications for therapy

Whether genes for drug resistance exist prior to infection or originate through mutation after infection is critical for our choice of treatment strategies. If the genes are already in the infecting population, strong, prolonged treatment with antibiotics rapidly and effectively selects for resistance. It kills the susceptible cells, thus eliminating their competition with the resistant cells, which are free to flourish. If those resistance genes are not already in the infecting population but arise through mutation only after the infection is established, antibiotic treatment greatly reduces the size of the infecting bacterial population and with it the probability that a gene will mutate into a resistant state. Most traditional recommendations for antibiotic therapy—hit them hard and finish the prescription—assume the second situation, but much evidence points to the high frequency of the first. That suggests that we could often slow the evolution of resistance by using only enough antibiotics to control the infection, buying time for the immune system to eliminate it without eliciting the evolution of resistance (Huijben et al. 2013).

Why does resistance evolve quickly and spread widely?

Antibiotic resistance evolves whenever antibiotics are used, and they are often used for excellent reasons: they can reduce suffering and save lives.

Antibiotics, however, are overused, misused, and used as growth promoters in agriculture to increase profits by people who are outsourcing the indirect costs.

Antibiotics do not cure most colds, coughs, and sore throats, which are caused by viruses, not by bacteria, yet many physicians, particularly pediatricians under pressure from parents to "do something" for their sick children, continue to write prescriptions for diseases caused by viruses. It is estimated that in the United States, there are 18 million antibiotic prescriptions written each year for the common cold, none of which are useful; 16 million for bronchitis, of which only 20% are useful; 13 million each for sore throat and sinusitis, of which only half are useful; and 23 million for ear infections, of which 70% are useful (Mellon, Benbrook, and Benbrook 2001). Antibiotics are often inappropriately prescribed.

Another major driver of resistance evolution is antibiotic use in agriculture. In 2001, Mellon, Benbrook, and Benbrook estimated that half of antibiotic use in the United States was agricultural, of which only 20% was therapeutic; the other 80% was used to promote growth in healthy animals. The antibiotics used to promote growth amounted to 25 million pounds per year added to animal feed and resulted in continually administered, subtherapeutic doses, the ideal regimen to elicit the evolution of antibiotic resistance. Resistant bacteria in farm animals are present in the processed meat and move through supermarket food chains onto dinner tables.

How can we delay, avoid, or prevent the evolution of antibiotic resistance?

We can reduce the rate of infection by cooking eggs and meat thoroughly and by washing our hands to slow transmission among humans. We can limit the use of antibacterial soaps and cleaners, which have little benefit beyond that already provided by plain soap. We can avoid prescribing antibiotics for viral infections, which will encounter resistance from patients, and we can eliminate antibiotic use in animal feed, which will elicit resistance in farmers, who profit from rapid growth of farm animals. These measures will all help, but not everyone will adhere to them, and we will still need to use antibiotics appropriately to control bacterial infections in patients. Antibiotic resistance will continue to evolve.

We can also modify existing drugs or discover new drugs. However, we have found few new classes of drugs recently. The costs of research and testing are enormous; many new drug candidates are failing because of toxicity issues; and history suggests, and experience confirms, that resistance will rapidly evolve every time a new drug is introduced. We need evolutionary insights to come up with therapies that are evolution-proof. Some initial ideas are discussed in the next section.

■ SUMMARY

Microbial pathogens rapidly evolve resistance, especially in emergency rooms, intensive care units, and agricultural settings. Nosocomial infections of multiply-resistant pathogens now kill more patients each year in the United States than do breast cancer and AIDS combined. The global cost of treating infections of multiply-resistant bacteria is on the order of $1 trillion/year. The genes for resistance often evolved long ago and far from any interaction with human drugs; they frequently move into human pathogens through horizontal gene transfer. Whether such genes are in the infecting bacterial population or originate through mutation after infection makes a critical difference to treatment strategy. Every new drug has become ineffective within a few years of its introduction, and the pipeline of drugs in development is drying up. All that focuses attention on ways of treating infections that slow or prevent an evolutionary counterresponse.

Therapies that mitigate evolutionary consequences

Currently, we have three options for therapies that could slow or avoid the evolution of antibiotic resistance. The first is to continue to use the antibiotics that we have but apply them in smaller doses for shorter periods. Here the logic is sound, but the method has not yet been tested in humans. The second is phage therapy, which uses viruses that specialize on bacteria. This approach has been tested in both model systems and humans, and it works. The third is to disrupt or destroy signals or substances produced by bacteria that are "public goods" in the sense that they benefit all members of the infecting bacterial population but impose a metabolic cost on the individual bacteria that produced them. This approach looks promising in a test on model organisms.

Slowing resistance evolution by managing therapy

Whether antibiotic therapy accelerates or retards resistance evolution critically depends on whether genetic variation for resistance is present in the population of infecting bacteria. If there are no resistance genes in the infecting population, antibiotic therapy strongly reduces the size of the bacterial population and with it the opportunities for mutations for resistance to arise. If some of the infecting bacteria carry genes for resistance, however, antibiotic therapy kills the susceptible strains but not the resistant ones, which then are free to multiply without having to compete with the strains that the antibiotics have eliminated. Unfortunately, most resistance genes evolved long before antibiotics were used, move frequently by horizontal gene transfer, and now exist at high frequency in the bacteria that live in emergency rooms and intensive care units. Thus, most bacterial infections—especially nosocomial infections—contain resistance genes, and antibiotic therapy often strongly and rapidly selects resistant strains.

   The rate at which resistance evolves in a given infection can be reduced by not using any more antibiotic than necessary to protect the patient. Lower doses used for shorter periods allow competing, susceptible strains to survive; they will suppress the evolution of the resistant strains. If such therapy can be properly managed so that patients are not endangered, it can give the immune system time to clear the infection without provoking resistance evolution (Read, Day, and Huijben 2011).

Phage therapy

Bacteriophage—viruses that specialize on infecting and killing bacteria—have several advantages over traditional antibiotics. Because they multiply exponentially on their bacterial resource, they persist in the patient's body while automatically adjusting the therapeutic dose: often only one dose is needed. Phage also mutate and evolve as rapidly as their bacterial hosts, producing descendants that can continue to recognize, infect, and kill. Whereas traditional antibiotics also kill useful gut bacteria that produce vitamins, help digest food, and maintain a healthy gut free of diarrhea, phage can target specific pathogenic bacteria without harming beneficial bacteria.

Phage were discovered in 1915 and have been used for therapy since 1926, mostly in Eastern Europe. Since the evolution of antibiotic resistance has become a serious problem, phage therapy has been increasingly viewed as a viable option worldwide (Figure 5.12).

There is some natural resistance to inoculating a human with a virus, even if that virus specializes on prokaryotes and cannot infect eukaryotic cells. Is phage therapy safe? Studies on mice have shown that phage can be as effective in treating bacterial infections as are antibiotics, and phage have recently been approved for protecting humans from beef contaminated with a pathogenic strain of *Escherichia coli* (O157:H7). Gut infections of multiply-resistant *Clostridium difficile* resulting in serious diarrhea are a frequent consequence of antibiotic use in hospitals. A recent test in a model of a human colon showed that therapy with the phage ΦCD27 significantly reduced the size of the pathogenic bacterial population and the amount of bacterial toxins without detrimental effects on commensal bacteria (Meader et al. 2013).

Phage therapy has some compelling advantages (Loc-Carrillo and Abedon 2011). Phage kill bacteria with no opportunity for recovery, they automatically adjust dosage, they have low inherent toxicity, they minimally disrupt commensals, there is no evolution of resistance, they can be rapidly discovered and flexibly applied, they can clear biofilms that resist antibiotics, they have the potential to be administered in a single dose, they have the potential to transmit protection, they have low environmental impact, and they are inexpensive to produce. There are also disadvantages. Their narrow host ranges make them inappropriate for broad-spectrum use on infections of unidentified bacteria, they can evolve, and they can interact with the immune system.

Therapy with viruses is not limited to phage attacking bacteria. In principle, it can be extended to nonpathogenic viruses capable of attacking eukaryotic cells and deployed to combat cancer without causing the evolution of resistance to chemotherapy. Preliminary trials on mice with viruses designed by artificial selection for that purpose have been promising.

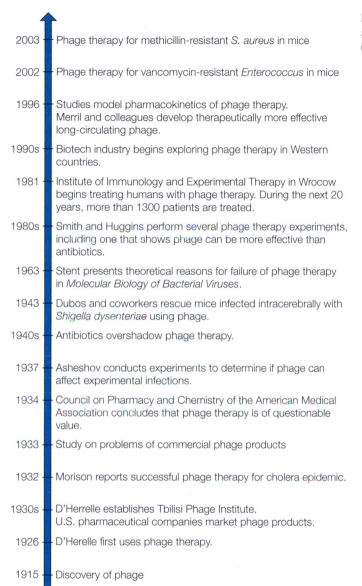

2003 — Phage therapy for methicillin-resistant *S. aureus* in mice

2002 — Phage therapy for vancomycin-resistant *Enterococcus* in mice

1996 — Studies model pharmacokinetics of phage therapy. Merril and colleagues develop therapeutically more effective long-circulating phage.

1990s — Biotech industry begins exploring phage therapy in Western countries.

1981 — Institute of Immunology and Experimental Therapy in Wrocow begins treating humans with phage therapy. During the next 20 years, more than 1300 patients are treated.

1980s — Smith and Huggins perform several phage therapy experiments, including one that shows phage can be more effective than antibiotics.

1963 — Stent presents theoretical reasons for failure of phage therapy in *Molecular Biology of Bacterial Viruses*.

1943 — Dubos and coworkers rescue mice infected intracerebrally with *Shigella dysenteriae* using phage.

1940s — Antibiotics overshadow phage therapy.

1937 — Asheshov conducts experiments to determine if phage can affect experimental infections.

1934 — Council on Pharmacy and Chemistry of the American Medical Association concludes that phage therapy is of questionable value.

1933 — Study on problems of commercial phage products

1932 — Morison reports successful phage therapy for cholera epidemic.

1930s — D'Herrelle establishes Tbilisi Phage Institute. U.S. pharmaceutical companies market phage products.

1926 — D'Herelle first uses phage therapy.

1915 — Discovery of phage

Figure 5.12   Phage therapy has a long history and some success. (After Levin and Bull 2004.)

Disrupting bacterial production of public goods

A public good is something produced at a cost by an individual that benefits the population as a whole. Bacteria produce at least two types of public goods: molecules that signal the density of the local population and siderophores that are produced to scavenge iron. Both are virulence factors: bacteria will not attack host tissue until they sense that enough of them are present to have a good chance of overcoming the host immune system, and they need iron to grow and reproduce.

Extracellular quenching of shared virulence factors has at least three advantages over conventional antibiotics. First, its extracellular action avoids any resistance mechanisms that depend on blocking the entry of a drug to or speeding its exit from the bacterial cell. Second, any fitness benefits experienced by emerging mutants are diluted across all the cells that are interacting locally. Third, some of the agents used in such quenching are not likely to have been experienced previously by bacterial populations, which are therefore unlikely to carry evolved resistance genes.

This idea has been tested on pathogenic *Pseudomona aeruginosa* bacteria infecting caterpillar hosts, in which it was contrasted with conventional antibiotic treatments. Gallium treatments quenched iron-scavenging siderophores, attenuated bacterial virulence and growth, and retained efficacy over the period during which conventional antibiotic treatments caused the evolution of resistance (Ross-Gillespie et al. 2014). It thus appears that extracellular quenching of bacterial public goods may offer an evolution-proof alternative to antibiotic therapy.

■ SUMMARY

Because pathogens rapidly evolve resistance to drugs, at least three therapeutic strategies with less serious evolved consequences are being explored. Evolution of resistance can be slowed by treating less intensely and for shorter periods with conventional antibiotics, phage therapy has proven effective in some settings and deserves wider consideration and testing, and extracellular quenching of public goods can disrupt bacterial growth and survival.

Cancer

In recent years, evolutionary insights have revolutionized thinking about cancer in several ways. Events deep in our evolutionary history traded advantages for disadvantages that have left us particularly susceptible to the origin and spread of cancer cells. When a cancer does originate, it undergoes a process of clonal selection, a dynamic evolutionary process fueled by genetic heterogeneity with crucial implications for metastasis and chemotherapy. In so doing, it creates its own unique evolutionary history within each patient, a history that can now be usefully reconstructed with molecular phylogenetics.

Cancer basics

We begin this chapter by describing the types of cancers, their hallmarks, their prevalence, and the reasons humans have more cancer than other species. Cancer is a family of many diseases, not a single disease, whose members share some important characteristics while differing in others.

Types of cancer

We start with some definitions. A tumor is any abnormal swelling of the flesh; tumors are a broader category than cancers. Cancers are a group of diseases characterized by uncontrolled growth and spread of abnormal cells. Not all masses of abnormal cells—neoplasms—are threatening. Benign neoplasms, such as skin moles, are localized and remain so. Premalignant neoplasms, such as colon polyps, do not invade and destroy, but given time, they can develop that capacity. Malignant neoplasms undergo metastasis, producing cells that can disperse, invade, and colonize other tissues. They are the life-threatening cancers.

Cancers can be classified by tissue type and by cell type. Carcinoma, sarcoma, and leukemia refer to classifications by tissue type. Carcinomas develop in epithelial cells, represent 90% of all tumors, and are derived mostly from ectoderm; some are derived from endoderm. Sarcomas develop in connective tissue, represent 2% of all tumors, and are derived from mesoderm. Leukemias develop in circulatory or lymphatic tissue, represent 8% of all tumors, and are derived from mesoderm. Cancers are also classified by cell type. Squamous refers to flat epithelial cells, myeloid to granulocytes and macrophages, lymphoid to lymphocytes or macrophages, and adenomatous to ductal or glandular cells (e.g., in the colon, prostate, or thyroid).

Hallmarks of cancer

Cancer cells have slipped out of cell cycle control, escaping the developmental monitoring systems that stabilize the differentiation of normal tissue. They no longer need external signals to stimulate proliferation, and they ignore signals to shut down proliferation or to initiate programmed cell death (apoptosis). Because they often have enzymes that stabilize their telomeres, they can divide indefinitely. They experience increased rates of mutation both for chromosomal abnormalities and for point mutations in their DNA sequences; these mutations produce a great diversity of clonal lineages, each with different genetic characteristics. The cancer cells that survive and proliferate have adaptations for evading destruction by the immune system. Tumors are not simply bags of cancer cells. They acquire many characteristics of normal tissues, such as recruitment of stromal cells, establishment of an extracellular matrix, and induction of angiogenesis: the growth of new blood vessels. In the late stages of cancer growth, clones arise that are capable of metastasis; they colonize other parts of the body.

Cancer prevalence

The lifetime risk of being diagnosed with a cancer of any type in a developed country that has gone through the demographic transition, which dramatically extended life span (see Chapter 10), is about 33%. In those countries, women experience a lifetime risk of breast cancer of about 12%. Autopsies on individuals who died of other causes reveal many covert malignant cancers and premalignant neoplasms in the prostate, breast, kidney, and thyroid as well as in other sites. All adults have thousands of precancerous neoplasms that have acquired some but not all mutations necessary for complete transformation to cancer. As those cells continue to divide, some acquire additional mutations that allow them to evade and ignore the remaining control mechanisms, including immune surveillance. As a result, if they lived long enough, all men would probably get prostate cancer, and all women would probably get breast cancer.

In 2006, about 720,000 men and 680,000 women died in the United States from cancer (Figure 6.1). The leading causes were lung and bronchial, then breast, prostate, colon and rectal, ovarian, pancreatic, and leukemic

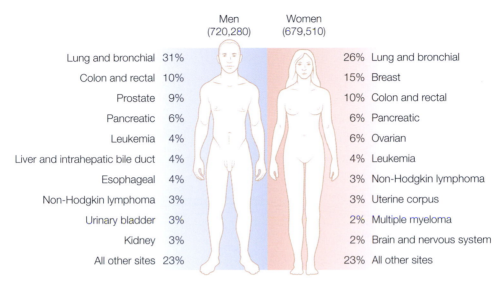

Figure 6.1   Deaths from cancer in the United States in 2006. Excludes basal and squamous-cell skin cancers and in situ carcinomas except urinary bladder. (After American Cancer Society 2006.)

cancers. In all these tissues cell division is frequent; recall that mutations occur during cell divisions.

Since 1930, great progress has been made in preventing deaths from heart disease, cerebrovascular diseases (strokes), and pneumonia and influenza. In contrast, there has been very little decrease in the rate of death from cancer. As we have discovered better treatments for the other major degenerative diseases, we have been prolonging the lives of people who might then die of cancer. Success in treatment has differed greatly with the type of cancer (Figure 6.2), with striking progress for stomach and uterine cancers; some progress for colon, rectum, and breast cancers; but little progress for pancreatic and liver cancers. Therapies developed since 2003 have improved survival for some patients with leukemia and melanoma. The trends in mortality from lung and bronchial cancers reflect the impact of smoking, which started to decrease for men before it did for women.

Why humans are especially susceptible to cancer

There are at least four reasons humans have more cancer than other species. First, we now have a significant period of postreproductive life during which we are not directly contributing genes to future generations; it is not very important to natural selection. Deaths occurring late in life have little impact on reproductive success, and evolution will not prioritize measures that might reduce such risks.

Second, we have highly invasive placentas produced by embryonic stem cells whose ability to move through tissue and establish themselves in new

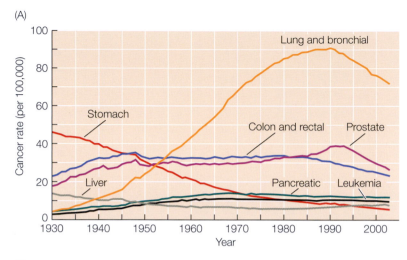

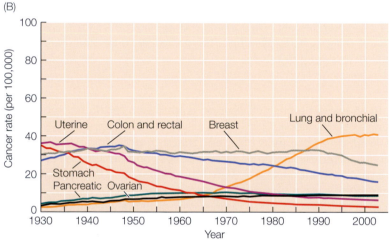

Figure 6.2 Risk of death by cancer for men (A) and women (B) in the United States from 1930 to 2003. Age-adjusted to the 2000 U.S. standard population. (After Wingo et al. 2003.)

sites, such as maternal endometrium and spiral arteries, preadapts them for metastasis. Every cell in the body contains the genetic information needed for tissue invasion. Under normal circumstances, it is not expressed, but if mutations occur in the genes that control the cell cycle and tissue differentiation, it can be expressed in some of the cancer cells, enabling their motility and metastasis.

Third, our biology is mismatched to risk factors produced by civilization, including tobacco; alcohol; a high-calorie, high-fat diet containing refined sugar and grain; air and water pollution; and contraceptives. Our development and physiology have had no evolutionary experience of these factors.

Fourth, some of our reproductive cancers are in part the by-products of our unique sexuality. We have continuous cycling, continuous receptivity, and the potential for continuous activity through the central decades of our lives. Those are all traits that are made possible by continuous tissue proliferation and cell division that increase the risk of mutation.

■ SUMMARY

Cancer is a disease caused by inappropriate and uncontrolled cell growth and movement. It is a collection of diseases, not a single disease, that is an increasingly important source of mortality in developed countries. If we lived long enough because we had escaped death for other reasons, we would all probably die of cancer. Our ability to treat cancer has not improved nearly as much as our ability to treat other diseases.

Why we are susceptible

In addition to the four reasons why humans are particularly susceptible to cancer, there are three basic evolutionary reasons why animals in general get cancer at all. The first has to do with the evolution of aging: maintenance will be neglected if reproduction can be sufficiently improved through that neglect. The second is a condition shared by all multicellular animals: we need stem cells for tissue renewal and repair, but they are pre-adapted for the cancerous lifestyle, information about which is therefore carried in every cell nucleus in the body. The third is a condition that follows directly from our body mass and is experienced much more strongly by larger animals like elephants and whales: the number of somatic mutations that occur during our development is overwhelmingly large, generating a universe of possibilities so great that events that are improbable in a single cell become very likely in an entire body. We now explore these conditions in more detail.

The causes and consequences of aging

As we saw in Chapter 2, aging is a by-product of selection for reproductive performance earlier in life. Maintenance is not perfect because the neglect of some maintenance can release resources that improve reproduction. This situation arises for two reasons, one having to do with selection, the other with genetics. Selection on the aged will always be weaker than selection on the young simply because mortality that does not necessarily have anything to do with aging ensures that there will always be fewer old people than young people. As a consequence, any mutation that has a sufficiently positive effect early in life but some negative effect later in life will be able to invade the population because in the evolutionary currency of gene frequencies, its benefits will outweigh its costs.

Prior to the demographic transition, such genes could accumulate in the population to a greater extent than is currently possible because they carried costs of mortality in the aged that often were not paid because people died for other reasons—primarily infectious disease and childbirth—before they got that old. When modern civilization introduced clean water supplies, vaccination, antibiotics, and all the other recent medical advances, it greatly reduced mortality from infectious disease and childbirth. Cardiovascular disease emerged as a major killer, but when we discovered the risks posed by smoking, obesity, and lack of exercise and began to prescribe statins and aspirin, the deaths caused by heart attacks and strokes started to decline in frequency, allowing more people to survive to ages when the risks of cancer and dementia are high. That recent development continues to unfold across the globe.

The multicellular covenant and the double edged sword of stem cells

The second condition is ancient. Multicellularity originated in animals about a billion years ago. It was stabilized by a fundamental covenant among cells: somatic cells stopped reproducing, germ cells propagated the genes, and stem cells repaired tissues. That led to a division of labor expressed in the evolution of different tissue types and stabilized by developmental mechanisms, both epigenetic and immune. Cancer breaks that covenant.

Most cancers originate in stem cells, a great innovation that allows the maintenance of tissues. Stem cells divide slowly and retain the potential to differentiate, some globally (embryonic stem cells), others locally (stem cells located in specific tissues). There are stem cells positioned all over the body, ready to replace cells that wear out and are discarded, particularly in lungs, the intestine, skin, and bone marrow. The self-renewal capacity of stem cells and their potential to differentiate preadapt them to a cancerous lifestyle.

Most of the time, somatic cells are kept under control. Throughout the body, tumor-suppressor genes control differentiation and deal with damage. If cells start to proliferate when they should not, they receive a signal to commit suicide (apoptosis). If DNA is damaged, attempts are made to repair it, but if the damage cannot be repaired, the cell receives a signal to apoptose. Both responses are controlled by the tumor-suppressor gene *p53*, which is disabled by mutations in most cancers. Several somatic mutations have to occur in human cells to transform them into cancer cells. These mutations affect cellular characteristics that roughly correspond to the hallmarks of cancer mentioned above.

Somatic mutations occur for several reasons. Perhaps most fundamental is that evolution has adjusted the mutation rate so that it is low but still appreciable. That is because mutation is the ultimate source of the genetic variation that provides the fuel for selection. Any lineage in which the mutation rate could be adjusted to zero—even if that were physically and chemically possible, which it probably is not—would go extinct because it could not keep pace with a changing environment. Thus, the mutation rate per nucleotide per cell division is low but finite, about 10^{-9}, and 3.3

billion nucleotides have to be copied every time a human cell divides. Mutations also occur because DNA experiences damage both from environmental factors, like ultraviolet light and tobacco smoke, and from the free radicals released by inflammation. In most cases, chronic exposures with accumulating effects are much worse than acute exposures. Finally, inactivation of genes such as *p53*, which are involved in quality control of the genome, leads to genomic instability and greatly increases the mutation rate in cancer cells.

Are there really genes "for" cancer?

Most genes involved in cancer function in early embryogenesis, mediating cell proliferation, survival, adhesion, and migration. Some of the genes that perform these functions in early development are silenced during the terminal differentiation of most tissues. Oncogenic mutations disrupt the epigenetic controls on these genes, allowing them to express normal cell functions in developmentally inappropriate contexts. Such genes are then expressing antagonistic pleiotropy: a benefit at an early age connected to a cost at a late age. If natural selection were to protect us from cancer by eliminating these genes, it would pay too large a cost early in development. Instead of eliminating them, selection has reinforced controls and provided mechanisms to detect and then correct or eliminate cells with precancerous mutations. People who have inherited a predisposition to cancer often have deficiencies in the mechanisms that stabilize and control tissue differentiation and that detect and correct cellular damage (Ellison, pers. comm. 2012).

Somatic mutations and cancer: Heterogeneous risk

Not all somatic mutations are equally dangerous; only about 1% of the genes in the genome—roughly 350—are involved in oncogenesis. Most of the dangerous mutations occur in stem cells, not in differentiated cells. It is mutations that occur early in development that are most important for cancer because they produce a subsequent lineage of many cells in which other mutations can accumulate. Each cell lineage in the body thus develops its own unique history. With more than 10^{16} cells per individual per lifetime, each of us contains a history of cell lineages that is greater than the history of all individual humans who have ever lived. Thus, the opportunities for dangerous somatic mutations are immense.

Some tissues, such as the epithelia of lungs, the intestine, and skin, are directly exposed to environmental carcinogens. Some tissues are also more susceptible to cancer than others because they contain lineages of mitotically active cells, and each cycle of mitosis is an opportunity for mutation. Cancer arises more frequently in the endometrium than in fallopian tubes, for example, and in the secretory tissues of the gut, breast, cervix, and prostate than in the associated muscles. One reason that children are more susceptible to brain and bone cancers than are adults is that brain and bone tissues are growing rapidly in children but not in adults. When cancers do appear in tissues that are not mitotically active, they are likely to be due to mutations that cause cell dedifferentiation and acquisition of stem-cell characteristics.

■ SUMMARY

Because we age as a by-product of selection for reproductive performance, maintenance is neglected in older organisms for good evolutionary reasons. We need stem cells to maintain our tissues, but mutations in stem cells can uncover characteristics that predispose them to be cancerous. Although we need a modest, low rate of mutation to maintain evolution, the opportunities for somatic mutation are immense, and even individually improbable events become highly probable when we consider the large number of cells in which they can occur. We need cell functions early in development that support cancer later in development if not properly controlled. Humans are especially vulnerable to cancer because we have a long postreproductive period that is relatively invisible to selection, we are exposed to novel risk factors not previously encountered in evolution, and we have unusual sexuality. In sum, we trade vulnerability to cancer for valuable benefits in reproduction, tissue maintenance, evolvability, development, and sexuality.

Every cancer is an instance of clonal evolution

Cancer evolution is short sighted and selfish, playing out within the lifetime of the individual affected. Because human cancers are not transmitted to other individuals, they die with the person affected; thus, they cannot accumulate a sequential history of progressive adaptations. Cancers do evolve, and they evolve rapidly, but their evolution is rough and ready, operating on cell clones for about 40 to 50 cell generations. They never achieve a state of finely polished design, but they are able to acquire the ability to evade or ignore the immune system, and, to a certain degree, metastases can adapt to the tissues they invade.

Cancer clones originate through mutations and then compete for space and resources. Two of the key genes involved in cancer are *p16* and *p53*; both control the cell cycle and DNA repair. Damage to genes controlling DNA repair increases the mutation rate, generating the genetic variation among clones that fuels responses to selection. That connection is well demonstrated by studies showing that the greater the genetic variation in a tumor, the more likely it will progress to malignancy.

Somatic cell evolution within an individual reflects in a single lifetime the branching process of genetic evolution in the germline that occurs over many generations (Figure 6.3). Somatic cells can differ genetically through point mutations, epigenetically through changes in patterns of gene silencing, and cytogenetically through chromosomal mutations, duplications, and deletions. Those somatic mutations result in a mosaic genetic pattern unique to each individual. Most of them do not result in cancer.

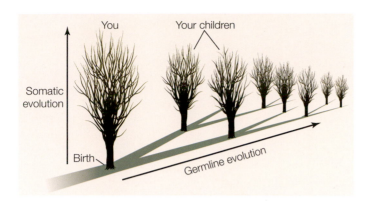

Figure 6.3 Somatic evolution occurs orthogonally to germline evolution. (After C. Nunn, pers. comm. 2012)

Neoplasms evolve by natural selection

Cancer is produced by competition among the clonal lineages that are generated by somatic mutation. Those somatic mutations produce variation among clones that affects the reproduction and survival of the cells. It is now well known that cancers display extreme genetic heterogeneity. That heterogeneity has been detected by cytogenetics, by karyotyping, by comparative genomic hybridization, by fluorescent in situ hybridization, as loss of heterozygosity, as shifts in microsatellite patterns, as changes in methylation patterns, and by DNA sequencing. For example, in a sample of lung cancer tissue, a pathologist can stain for particular chromosomes and observe one to sixteen copies of chromosome 17 and one to three copies of chromosome 11; the number of each in normal tissue is two.

Mutations in two genes are found particularly frequently in cancer clones and can lead to a high risk of cancer in offspring if they occur in the germline of parents. The *CDKN2A* gene produces at least two transcripts from alternative promoters. One encodes p16^{INK4A}, an inhibitor of cylin-dependent kinases CDK4 and CDK6, which are important for cell cycle initiation. The other transcript encodes p14^{ARF}, a protein that stabilizes the transcription factor p53, produced by the *TP53* gene, by sequestering MDM2, a protein that degrades p53. The p53 transcription factor regulates the cell cycle, suppressing tumor formation. It can activate DNA repair proteins, arrest the cell cycle when it recognizes DNA damage, and initiate apoptosis if the DNA damage is too serious to be repaired. People who inherit only one functional copy of the *TP53* gene are at high risk of a disorder known as Li-Fraumeni syndrome, which increases the chances of cancer developing in early adulthood. More than half of human cancers contain a mutation or deletion of the *TP53* gene. Clonal expansions are signatures of natural selection that record the reproductive success of cells in those clones. Clonal expansions of *p16* and *p53* lesions have been found in cancers of the skin, lung, breast, head, neck, bladder, brain, kidney, prostate, colon, stomach, and esophagus.

The relationship between the genetic heterogeneity of a premalignant tumor and the probability that it will progress to malignancy has been studied in Barrett's esophagus, a condition that predisposes for malignant

Figure 6.4 The genetic heterogeneity of esophageal tissue in patients with Barrett's esophagus helped predict which patients would progress to malignant cancer. Incidence of adenocarcinomas is a proxy for likelihood of developing malignant cancer. The upper quartile represents cases with higher numbers of clones (based on loss of heterozygosity); these cases were significantly more likely to develop into malignant cancer than those in the other three quartiles. (After Maley et al. 2006.)

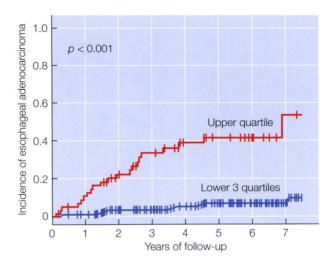

esophageal adenocarcinoma (Maley et al. 2006). Three measures of diversity were used: the number of different clones (Figure 6.4); the Shannon index of diversity, which measures both the number and the relative abundance of the different mutants; and the genetic divergence of the mutants, that is, how different they were from one another. All three measures helped predict which patients would develop cancer.

Clonal evolution has some special features

In clonal evolution, clones have to acquire mutations sequentially because they do not exchange genetic information sexually. Here the sequence of mutations is critical because a clone bearing one mutation or set of mutations may be able to outcompete another clone that might have a precursor of the set of mutations that could eventually beat all opponents (Figure 6.5).

Figure 6.5A communicates the idea that in clonal evolution, a set of mutations that in combination could beat all opponents (ABC) must be accumulated sequentially. During that period of sequential accumulation, one clone can suppress the growth of other clones. In this case, clone A suppresses the growth of clone AC, but clone AB then outcompetes clone A. Eventually, the C mutation arises in cells of clone AB, and then clone ABC competitively replaces clone AB. Figure 6.5B shows how that idea plays out in the specific case of Barrett's esophagus. Clones with a single copy of mutant *p16* spread through normal tissue and suppress clones with a single copy of mutant *p53*. Clones with two copies of mutant *p16* then replace both single-copy mutants. When a mutation to *p53* then arises in the cells that have two copies of mutant *p16*, the resulting clone outcompetes all opponents and progresses to malignant cancer. This view of cancer as a process of dynamic clonal evolution has now become standard in oncology (Figure 6.6).

(A) Clonal evolution in general

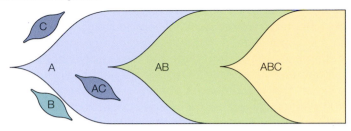

(B) Clonal evolution in Barrett's esophagus

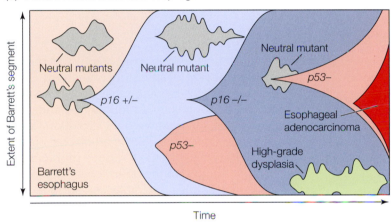

Figure 6.5   (A) Clonal evolution in cancer mirrors the general principles of clonal evolution anywhere. (B) Genes *p16* and *p53* are tumor suppressors; (+) indicates normal function, (–) indicates a mutation causing loss of function. High-grade dysplasia is a severe precancerous tissue change. (After Crespi and Summers 2005.)

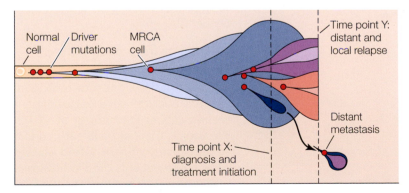

Figure 6.6   The clonal evolution of cancer starts long before diagnosis and treatment. Cells that initiate clonal expansions (most recent common ancestors, MRCAs) contain mutations that are considered to be drivers (as well as other mutations). (After Yates and Campbell 2012.)

■ SUMMARY

Cancer starts when a combination of mutations—some somatic, some possibly inherited from the germline—produces cells that first slip out of cell cycle and cell death controls, growing unconstrained, and then evolve the ability to move through the circulatory system to establish new colonies in distant tissues. Cancers are genetically heterogeneous sets of clones that compete with one another for resources and space. That competition generates natural selection that favors clones that can grow fast, disperse, and invade. Such evolution is short sighted and selfish because the cancer dies with the patient; it is not transmitted to the offspring. The adaptations it can produce are limited and rough.

Cancer phylogenetics

The techniques of modern molecular phylogenetics can reconstruct the history of a cancer within an individual patient. That history often turns out to have been longer than expected: the index mutation that started a cancer clone may have occurred decades earlier in a much younger person. Cancer phylogenies are starting to reveal some convergent evolution, which lends some hope to the search for general characteristics on which therapies could focus, but they also reveal heterogeneities that pose problems for boutique therapies focused on single samples. The molecular phylogeography of brain tumors has revealed that multiple clones survive chemotherapy. Analysis of cancer phylogenetic trees has also revealed that therapy with nonsteroidal anti-inflammatory drugs (NSAIDs) reduced the rate of acquisition of somatic mutations, pointing toward one reason these drugs reduce cancer risk.

An insight from pancreatic cancer

For historical insight into individual cancers, consider seven cases of pancreatic cancer (Yachida et al. 2010). We usually think of pancreatic cancer as particularly virulent, killing quickly. After these patients died, metastases were harvested from the liver and lung and their DNA sequences were used to construct a phylogenetic tree on which the timing of the major events in the evolution of that cancer could be estimated (Figure 6.7). In these cases,

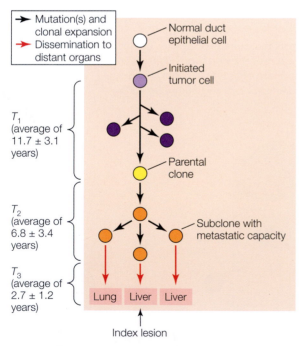

Figure 6.7 Distant metastasis occurred late in the evolution of a pancreatic cancer. (After Yachida et al. 2010.)

the tumors were initiated at least 18 years before the patients died. For about a decade, clonal evolution proceeded in the nonmetastatic primary tumors. Subclones with metastatic capacity evolved about 5–6 years before metastases established themselves and began to grow in the lung and liver. It was at that point that the cancers were diagnosed as index lesions in the liver, too late for effective therapy. Comparison of the DNA sequences of the metastases in the lung and liver revealed some evidence of limited local adaptation to those different tissue environments.

The problem of genetic heterogeneity

The simplest interpretation of clonal evolution is that one clone will win, eliminating the others. For at least two reasons, that is not what usually happens. First, there may not be enough time for the process to go to completion, and second, metastatic and blood cancers inhabit spatially diverse environments with many niches in which clones can hide from competition. Thus, many cancers remain genetically heterogeneous despite clonal evolution.

For example, detailed analysis of a case of acute lymphoblastic leukemia (K. Anderson et al. 2011) revealed that at least nine clones were present, including the ancestral clones, which had not been eliminated (Figure 6.8). No single clone had replaced the others, and the mutational history was marked by repeated loss of two genes (*PAX5* and *ETV6*).

A similar conclusion was reached in a study of renal-cell carcinoma in which the primary tumor on the kidney had sent metastases into the lung and chest wall (Gerlinger et al. 2012). Both the primary tumor and the metastases contained heterogeneous clones. This evolutionary divergence

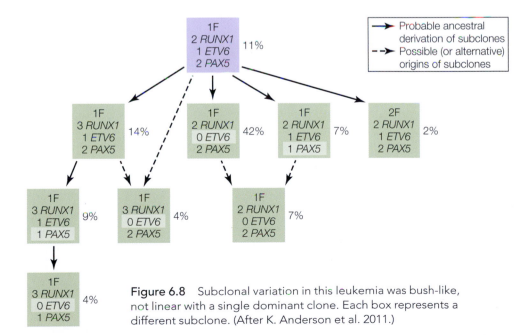

Figure 6.8 Subclonal variation in this leukemia was bush-like, not linear with a single dominant clone. Each box represents a different subclone. (After K. Anderson et al. 2011.)

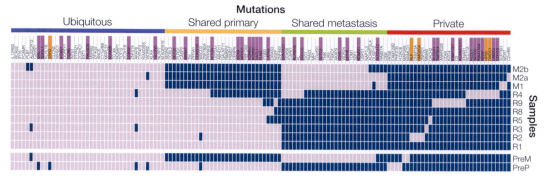

Figure 6.9   The distribution of mutations in a case of metastatic renal-cell carcinoma. Rows: samples (labeled on right). Columns: mutations. Gray, mutation present; blue, mutation absent; purple gene names, mutation validated; orange gene names, validation failed. Samples: PreP, pretreatment biopsy samples of the primary tumor; PreM, pretreatment biopsy samples of the chest-wall metastases; R1 to R9, primary-tumor regions of the nephrectomy specimen; M1, a perinephric metastasis in the nephrectomy specimen; M2a and M2b, two regions of the excised chest-wall metastasis. (After Gerlinger et al. 2012.)

while maintaining genetic diversity poses problems for therapies because therapies designed to target the primary tumor would miss variation in the metastases and therapies designed to target one metastasis would miss variation in the others. Some of the mutations were ubiquitous, probably having occurred in the last common ancestral cell that started the primary tumor. Of the others, some were shared across all sites in the primary tumor, some were shared among metastases, and still others were private, occurring only within a single metastasis (Figure 6.9).

The amount of adaptive evolution that had occurred could be estimated by looking at the number of nonsynonymous mutations along each branch in the phylogenetic tree (Figure 6.10). From this picture, we can draw two important conclusions. First, the genetic response to clonal selection as measured by the number of nonsynonymous mutations is a reality that underpins both growth and metastasis. Second, we need to treat the branches, not the twigs. To do so, we need to recognize them in time to save the patient.

The difficulty of that task is underlined by a study of a glioblastoma, a malignant brain cancer, in which multiple clones survived chemotherapy and each of them had different genes for resistance (Sottoriva et al. 2013). Unless treatments eliminated all surviving clones, survivors would persist and eventually flourish. A single biopsy would not give sufficient information to treat the entire tree, and missing a surviving clone could be fatal.

Using phylogenies to evaluate preventive therapies

NSAIDs are prescribed for many conditions, including heart disease, and some patients who use them develop cancer. Analyzing the phylogenetic

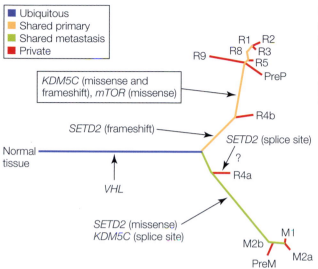

Figure 6.10 Phylogenetic relationships of the tumor samples described in Figure 6.9. The branch lengths are proportional to the number of nonsynonymous mutations; arrows indicate potential driver mutations. (After Gerlinger et al. 2012.)

histories of cancers in patients with a known history of NSAID use allows us to see whether the rate of clonal branching in the trees was greater before or after drug therapy was initiated (Kostadinov et al. 2013). That contrast showed a striking result (Figure 6.11). For patients whose Barrett's esophagus progressed to esophageal adenocarcinoma, taking NSAIDs reduced the rate of somatic genetic alterations (point mutations, chromosomal mutations, and copy number variants) by one order of magnitude, thus illustrating the power of phylogenetic analyses and emphasizing the important role of inflammation in creating a mutagenic environment that increases cancer risk.

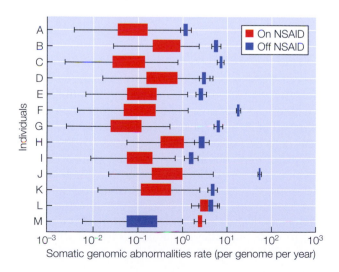

Figure 6.11 Nonsteroidal anti-inflammatory drugs (NSAIDs) reduce the rate of somatic mutations. (After Kostadinov et al. 2013.)

■ SUMMARY

The evidence for cancer as a process of clonal evolution is now over-whelming. Cancers originate earlier than had been expected, and if the parent clones can be detected, the time available for therapy is longer than usually envisioned. Single biopsies, however, seriously underestimate the clonal heterogeneity of both primary tumors and metastases. We need to develop methods of treating the entire tree, not a single twig or a single branch. Phylogenetic inference is proving useful for evaluating the efficacy of both preventive and therapeutic cancer treatments.

Immune evasion and suppression

Cancer cells share most of their genome with other cells in the body: they are modified self. The immune system contains many fail-safe mechanisms to ensure that it does not attack self. When the immune system does attack cancer, it does so because it can recognize self cells with pathological traits: inappropriate metabolism, growth, and movement. Cancer evolves to avoid or suppress immune attack by disrupting the cells, signals, and receptors deployed by the immune system. Those are the same options for evasion and suppression used by pathogens.

Most tumors are eliminated, some survive but are kept under control, and a few escape and progress to cancer. Elimination is done either through pathways intrinsic to the cell that are operated by tumor-repressor genes or by pathways extrinsic to the cell that involve recognition and killing by both the innate and the adaptive immune system. Some cancers escape elimination but remain small because they are kept under immune pressure: they are not clinically detectable. Escape and progression to cancer result either when the immune system is inhibited or when it is no longer effective due to mutations in cancer cells that alter their properties. Thus, the immune system is both a defense against cancer and one of the principle selective agents that shift the balance of competing clones toward those that can grow and metastasize.

Dealing with a major enemy: Natural killer cells

Natural killer (NK) cells are part of the innate immune system that can react quickly to pathogens and tumors with or without priming by antibodies. They have a repertoire of receptors that ensures self-tolerance while allowing interactions with infected cells and cancer cells. NK cells are thus a major problem for cancer cells because they activate when triggered by signals from stressed or transformed cells and then kill them, either directly through lysis or indirectly through the production of the cytokine IFNγ that promotes activation of T cells, macrophages, and dendritic cells (Figure 6.12). NK cells are covered with many receptors. Some are activating, and some are inhibitory. Others are chemotactic, and still others receive cytokine signals. There are also surface molecules involved in cell adhesion. These numerous, diverse receptors endow NK cells with flexible, calibrated

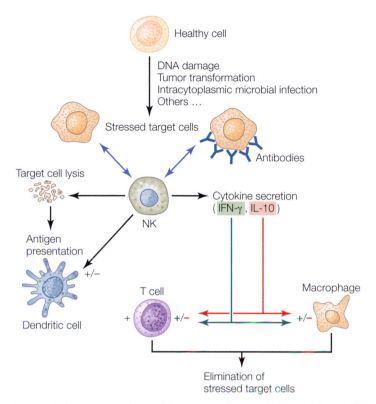

Figure 6.12 The biological functions of natural killer (NK) cells. Blue arrows indicate cellular signaling. (After Vivier et al. 2011.)

responses, enabling them to navigate appropriately between healthy self, cancerous self, and infected cells through a balance of activating and inhibiting signals (Figure 6.13). The ligands for activating receptors are expressed on cancer cells where they can be induced upon DNA damage. In contrast, the ligands for the inhibitory receptors (e.g., MHC class I, or MHC-I, molecules) are expressed on healthy, uninfected, and untransformed cells but are down-regulated on infected and transformed cells. Detection of a combination of signals for activating and inhibitory receptors allows the NK cells to distinguish between cells to eliminate and cells to spare.

However, these NK receptors also provide targets for disruption by cancers and pathogens. Analysis of a case of acute myeloid leukemia showed that the cancer had escaped surveillance by NK cells by disrupting their communication with dendritic cells and by inducing T_{Reg} cells to inhibit NK cells (Lion et al. 2012).

Exploiting a chink in the armor: Protection against autoimmunity

A specific population of T regulatory (T_{Reg}) cells suppresses autoimmune responses by inhibiting immune cells that would attack self. Tumor cells secrete a cytokine, TGF-β, that converts T cells into T_{Reg} cells. They thus

Figure 6.13 How natural killer cells integrate signals to determine response. (After Vivier et al. 2011.)

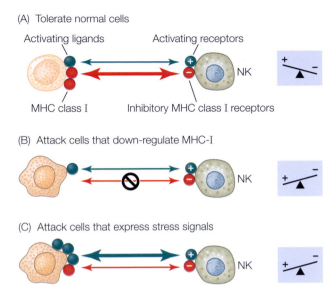

(A) Tolerate normal cells

Activating ligands   Activating receptors

NK

MHC class I   Inhibitory MHC class I receptors

(B) Attack cells that down-regulate MHC-I

NK

(C) Attack cells that express stress signals

NK

convert a population of T cells that would otherwise mediate attacks on tumors into T cells that inhibit immune attack. This conversion was shown by demonstrating that neutralizing the TGF-β derived from tumor cells blocked the conversion of T cells into T_{Reg} cells (Liu et al. 2007).

Tumor cells also down-regulate the expression of MHC-I molecules on their surface, but in so doing they are performing a balancing act: although that down-regulation helps them avoid attack by cytotoxic T cells, they must continue to express a low level of MHC-I to avoid attack by NK cells. Viruses manipulate the expression of MHC-I in a similar fashion (Villalba et al. 2013).

Modifying metabolism to create an advantage for cancer

Tumors modify their metabolism in ways that suppress cytotoxic and antigen-presenting cells (Kareva and Hahnfeldt 2013). Like other highly proliferative cells, cancer cells undergo aerobic glycolysis so that the majority of the glycolytic end product, pyruvate, is reduced to lactate instead of going into the TCA cycle. The excess of lactate secreted by tumor cells activates the hypoxia inducible factor (HIF-1α) in the nearby macrophages to induce production of vascular endothelial growth factor, a protein that stimulates angiogenesis, causing growth of the vasculature that supplies the tumor with oxygen and nutrients. In addition, lactate lowers the pH in the tumor microenvironment, which interferes with the maturation and activation of naive cytotoxic T cells and antigen-presenting cells. Thus, two hallmarks of cancer—aerobic glycolysis and immune suppression—may be mechanistically linked through a cascade of effects on the adaptive immune system (Figure 6.14). When cancer cells increase glycolysis and thereby lower pH, the production of interferon-γ (IFN-γ)—a cytokine produced by NK cells, cytotoxic T cells, and Th1 cells that activates macrophages and suppresses cancer cell proliferation—is decreased.

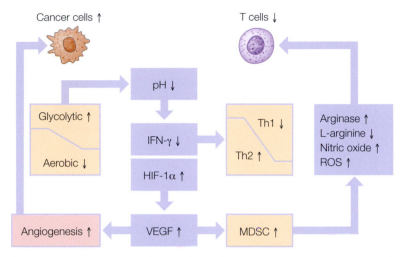

Figure 6.14   The up-regulation of glycolysis in tumors produces a cascade of effects on the cells of the adaptive immune system. IFN-γ, interferon γ; Th1, antitumor T helper cells; Th2, tumor-promoting T helper cells; HIF-1α, hypoxia-inducible factor α; VEGF, vascular endothelial growth factor; MDSC, myeloid-derived immuno-suppressive cells; ROS, reactive oxygen species. (After Kareva and Hahnfeldt 2013.)

■ SUMMARY

Most precancerous growths—the vast majority—are detected and destroyed by the immune system. Some remain but do not grow and cannot be detected by clinicians. Those that escape and progress to cancer have passed through a very strong selective filter, in the process acquiring the ability to suppress or evade immune surveillance. They do so by disabling NK cells, subverting the autoimmune response, down-regulating their expression of MHC-I molecules, and modifying the local microenvironment by changing their metabolism to interfere with the immune response. They thus resemble pathogens in manipulative capacity even though they have many fewer generations of selection to shape such adaptations.

Evolved resistance to chemotherapy

Chemotherapy strongly selects for the rapid evolution of resistant clones, just as antibiotic therapy rapidly selects antibiotic resistance in bacteria. As with antibiotic resistance, alternatives are now being explored that either slow the evolution of resistance or avoid it altogether. Those alternatives include targeted immunotherapy, adaptive therapy, and interventions that undermine the cooperation among cancer cells that is mediated by the production of public goods.

The efficacy of chemotherapy as a selective agent is a striking feature of the typical clinical experience of a cancer patient. If the patient arrives in the clinic with late-stage cancer that is metastasizing, surgery will often be considered ineffective; the physician will instead start a cytotoxic chemotherapy, usually once a week for three weeks. Such therapy attacks all proliferating cells in the body. The patient's hair falls out, the gut lining sloughs off, and the tumor shrinks, even disappearing from a magnetic resonance imaging (MRI) scan or an X-ray. (Note that a mass of 1 million cells is difficult to detect on an MRI scan.) A few months later, the tumor reappears. Using the same drug now has no effect because the surviving cancer clones are resistant to it. The physician tries a different drug, and the cycle repeats. Usually, the patient does not survive the third cycle. That is exactly how we would select for resistance in the laboratory.

The types of drugs used in chemotherapy usually fall into two categories: cytotoxins and targeted therapies. The cytotoxins include 5-flurouracil, which blocks DNA synthesis, and cisplatin, which binds to DNA, causing problems in DNA replication and transcription that trigger apoptosis. Targeted therapies include Gleevec, which inhibits the oncogene *BCR-Abl*; Tamoxifen, which blocks the estrogen receptor that is functional in breast cancer; and Gefinitib, which blocks the epithelial growth factor receptor. These therapies all select cells with resistance mutations. As a result, cancer treatment fails in late-stage tumors and especially in smokers, who have heightened mutation rates.

Targeted immunotherapy

Targeted immunotherapy exploits the existing capacity of the immune system to find and destroy cancer cells, attempting to enhance it by stimulating cytotoxic T cells and suppressing the immunosuppressive cells that increase up to 10-fold in cancer patients. Some therapies remove T cells from the patient's blood, select the clones that react specifically with the cancer, amplify them outside the body, and inject them back into the blood at greatly increased concentrations. There have been some dramatic successes as well as disappointing failures with extremely virulent cancers, including metastatic melanomas and ovarian cancer.

There are multiple immune signaling pathways modulating interactions in the tumor microenvironment (Figure 6.15). The most promising current approaches are based on the blockade of the inhibitory receptors CTLA4 and PD-1, which normally prevent autoimmune attacks. Blocking these inhibitory receptors unleashes T-cell responses against tumors, mimicking in many ways an autoimmune response.

Evolutionary approaches

One evolutionary approach that has worked with HIV is a drug cocktail, a combination of drugs that work through different mechanisms, each of which targets the product of a separate gene. Given that the probability of a mutation in one such gene is on the order of 10^{-6} per cell or viral generation, the probability that multiple genes would arise in a single cell is the product of the probabilities that each mutation would arise. For a three-drug cocktail, that would be 10^{-18}. This number is small enough to make the evolution of resistance highly unlikely because the number of

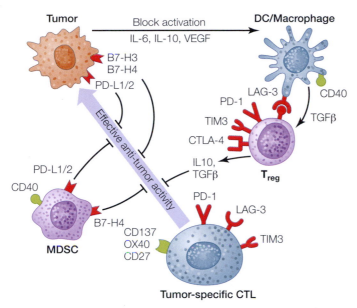

Figure 6.15 The signaling pathways between tumors and the immune system offer multiple opportunities for both the immune system and the tumor to suppress each other. DC, dendritic cell; CTLs, cytotoxic T lymphocytes; MDSC, myeloid-derived immunosuppressive cell. (After Pardoll and Drake 2012.)

cells or virions in a single patient is on the order of 10^{12} and the probability of a triple mutation in a population that size is $10^{12} \times 10^{-18} = 10^{-6}$, or about one in a million. Drug cocktails sound promising for cancer therapy until one realizes that the combined side effects of multiple drugs, each highly toxic, would be unsupportable, even fatal. They are not a promising approach for cancer because the cell functions they affect are present in both healthy cells and tumors.

Of the other ideas currently being explored, two have shown promise in preliminary tests. Adaptive therapy uses lower doses to control the cancer without eliminating the less aggressive clones that compete for resources with more aggressive clones. Another approach is targeting public goods to disrupt cooperation within the tumor, a manipulation that is difficult for clonal evolution to overcome.

Adaptive therapy works in mice (Figure 6.16). The supporting experiment contrasted the standard aggressive high-dose chemotherapy with lower doses that were adaptively adjusted to maintain constant tumor size. Mice receiving the standard high dose had tumors that rapidly evolved resistance, resulting in just as high a tumor burden as those receiving no treatment but with some delay. Those receiving adaptive therapy had tumors that did not increase in size. Adaptive therapy keeps the tumor from growing without killing it, thus avoiding the strong selection that elicits resistance. This strategy is the same one—"farm it, don't nuke it"—advocated to slow the evolution of antibiotic resistance (Read, Day, and Huijben 2011).

(A) Replicate 1

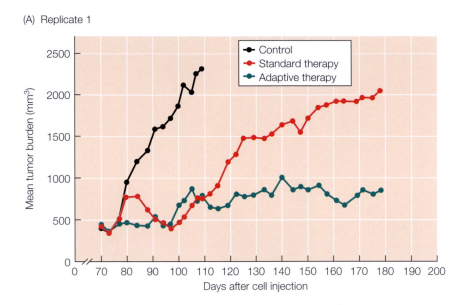

(B) Replicate 2

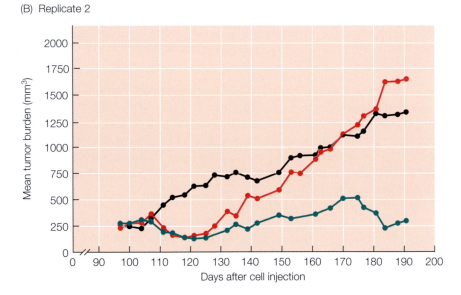

Figure 6.16 Two replicates of an experiment showing that adaptive therapy in mice maintains low tumor burden. (After Gatenby et al. 2009.)

The strategy of targeting public goods relies on an insight from game theory into the mechanisms that sustain cooperation (Archetti 2013). Cancer cells produce growth factors that suppress the immune system and stimulate angiogenesis. Doing so is costly in the sense that a cell that did not produce those growth factors could proliferate faster than a cell that did produce them. Thus, growth factors are a public good, and situations that are sustained by public goods are susceptible to invasion by selfish cells that do not pay the cost of producing them. This therapeutic manipulation removes cells from the cancer, modifies them by knocking out genes that code for growth factors, and then reinserts the modified cells into the cancer patient. The modified cells have a replication advantage and increase in frequency within the tumor, which eventually collapses for lack of essential growth factors. Preliminary experiments in cell culture have yielded promising results (Archetti, Ferraro, and Christofori 2015), but in vivo tests remain to be done.

The reason manipulation of public goods is a strategy against which it is difficult to evolve countermeasures is that once the population consists mostly of cells that are not producing the public good and therefore not paying the cost of producing it, it is very difficult for a mutant cell that does produce the public good to invade the population because its products are used by all, not only by itself and its descendants. It is creating an advantage for everyone but only paying the price itself. In principle, manipulations of public goods are not absolutely evolution-proof because evolution was, over long periods, able to evolve cooperation among cells, but they are certainly much more difficult to evolve around than are standard therapies.

■ SUMMARY

Resistance to chemotherapy evolves rapidly, and life span could be extended with alternative therapies. Targeted immunotherapy shows promise. Adaptive therapy has shown promise in mice and is moving into clinical trials in humans. Targeting public goods looks promising in theory and is being explored experimentally in model systems. Current cancer therapies are very expensive and therefore only available to those who can afford them, which has implications for health policy. Should access to health care depend on wealth? We will return to that issue in Chapter 10.

Reproductive Medicine

E volutionary insights help us understand many of the issues that
physicians confront in reproductive medicine. These insights are
particularly helpful in understanding menstruation, menopause,
the difficulty of childbirth, and diseases of pregnancy, such as
pre-eclampsia and pregnancy-related diabetes. There are also in-
triguing connections between invasive placentas and cancer (see
Chapter 6) and genomic conflicts over parental investment and
mental health (see Chapter 9), both cases in which reproductive
events early in life have consequences for later health. To set the
stage, we begin with the evolution of mammalian reproduction,
which has profoundly shaped the development of the female reproductive tract, the arena in which these events play out.

The evolution of mammalian reproduction

Mammals are endothermic amniotes distinguished from birds and
reptiles by having hair, three middle ear bones, mammary glands
in females, and a neocortex in the brain. The major groups of living mammals are the Prototheria (Monotremata; duck-billed platy-
puses and spiny echidnas) and the Theria, whose two groups are
the Metatheria (marsupials) and the Eutheria (placental mammals).
The first mammals, descendants of therapsids, appear in the fossil
record in the Late Triassic, about 225 mya; the Eutheria appear in
the Late Jurassic, about 160 mya.

Monotremes ("one hole") use the same orifice to defecate, uri-
nate, and reproduce; they lay leathery, uncalcified eggs; and their
females, which lack nipples, secrete milk from abdominal patches.
Marsupials have short gestation and give birth to relatively undeveloped offspring; in many, the offspring complete development in

a pouch on the mother's abdomen in which they suckle on a nipple. Placentals give birth to fully developed young and have longer gestation periods than marsupials.

The placental reproductive system is an evolutionary innovation. It derives from ancestral female organs that produced eggs that exited through a cloaca. The innovations—which include the uterus, the endometrium, the placenta, and the vagina—like all innovations, were accomplished by changes in the developmental control of cells and tissues. In this case, the genes used to control the new developmental fates of tissues in the reproductive tract had previously evolved to control the development of the body axis and limbs. A previously evolved, generalized toolkit was opportunistically applied in a new situation. Before we can understand how that toolkit was deployed to shape the placental reproductive tract, we need some background in evolutionary developmental genetics.

The evolutionary developmental genetics of multicellular eukaryotes

Development plays a key role in evolution, shaping the phenotypic variation that is available for selection. The developmental mechanisms that are shared among related organisms and that differ among major groups constrain responses to selection in ways shared by relatives. That is one reason species resemble shared ancestors in their basic body plans and ancestry influences responses to selection. There are other constraints on the developmental changes that underlie all evolutionary change in traits. Genes can only influence the construction of organisms through the properties of the materials that organisms use and the systems of chemical and biological interactions that they inherit. Because much of the phenotype is determined by the properties of those selected materials and those inherited systems, genes cannot directly influence every detail of the phenotype. In addition, because developmental systems are themselves inherited from ancestors, evolution has changed their properties by modifying ancestral systems. In so doing, it had to use whatever variants were available at the time that did a better job. Evolution does not design organisms a priori as would an engineer; rather, it works with inherited systems that contain constraints and trade-offs.

Development coordinates the fate of the many cell types in multicellular organisms. These cells communicate with one another, sending and receiving information on their relative positions and developmental states. That information is used to change cell fates. Because virtually all cells contain all the information needed to produce the entire organism, gene expression regulates development, turning some genes on and others off. Much of trait evolution has been mediated by regulatory changes in when and where similar sets of structural genes are expressed as well as by changes in the structural genes themselves.

The major players in multicellular eukaryotic gene regulation are the DNA-binding transcription factors and their coactivators and corepressors that interact to turn genes on or off (Figure 7.1). Upstream of the transcribed

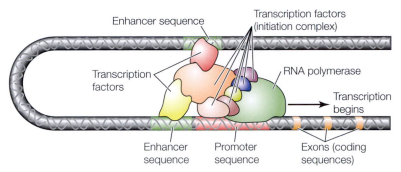

Figure 7.1   The major players in eukaryotic gene regulation. (After Gilbert and Epel 2015.)

region are enhancer and promoter sequences to which transcription factors bind, forming an initiation complex that stabilizes RNA polymerase as it binds to CpG islands or to the TATA-box sequence in the promoter region. It takes a specific combination of transcription factors to switch on the expression of a given gene; much of the specificity of expression in space and time is determined by that combinatorial control, which allows a relatively small set of transcription factor genes to be mixed and matched in many combinations to specify a relatively large number of tissue types.

In addition to DNA-binding transcription factors, gene expression is also controlled by covalent modifications of histone tails. The N-terminal sequences of histones can be modified with covalent bonds to several residues; those modifications include acetylation, methylation, and phosphorylation. Lysine acetylation and methylation play particularly important roles in establishing gene regulatory programs. For example, acetylation of the histone H3 at lysine 9 (H3K9) is associated with active chromatin, whereas methylation of the same lysine is associated with gene repression. Similarly, H3K4 methylation marks active genes, whereas H3K27 methylation is responsible for stable silencing of genes mediated by the polycomb repressive complex, yielding alternative cell fates such as muscle-specific genes in liver cells (hepatocytes). Transcription factors that can recruit histone-modifying enzymes often participate in cell fate decisions and are therefore referred to as master regulators. MyoD (myogenic differentiation), HNF4a (hepatocyte nuclear factor 4a), and PU.1 (proto-oncogene.1) are master regulators of myocyte, hepatocyte, and macrophage lineages, respectively.

When development starts, whether in the egg or in tissue primordia, concentration gradients of signaling molecules and transcriptional regulators yield positional information that initiates the organization of the body axis and limbs into anterior-posterior, dorsal-ventral, and proximal-distal fields. The genes that control the general pattern switch on first and are followed by the genes that control progressive levels of detail. Within the general areas set up early in development, transcription factors acting as combinations of activators and repressors define specific regions where precise subsets of genes are expressed. That combinatorial control permits

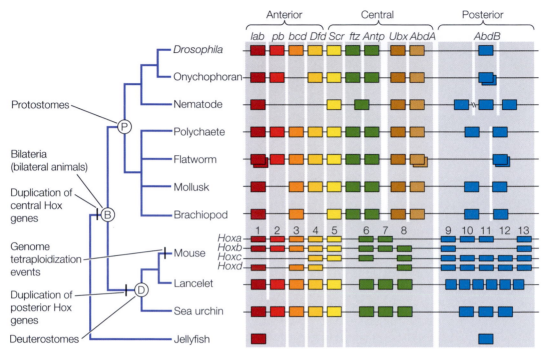

Figure 7.2 The Hox gene family is ancient, with deeply conserved sequence and function. (After Carroll, Grenier, and Weatherbee 2005; Janssen et al. 2014.)

a huge diversity of cell- and tissue-specific gene expression patterns. In genetic control networks, genes that produce transcription factors regulate other genes that also produce transcription factors. This regulatory flexibility allows genes that play one role in embryos to play other roles in adults.

In animals, the Hox, or homeobox, genes are a particularly important set of developmental control genes. They got that name because mutations in these genes changed the identity of major developmental modules, resulting in flies, for example, with eyes or antennae in abnormal locations. Such mutations are called homeotic. When the homeotic genes were sequenced, they were found to share a deeply conserved DNA sequence around which one could draw a box, hence "homeobox." That DNA sequence—which codes for an amino acid sequence that forms the DNA-binding portion of a transcription factor, a structure that has not changed in billions of years—is a signal in the wilderness of the genome that a gene for a transcription factor is present.

The Hox genes form a family of genes, initially produced by gene duplications, that share an ancient and deeply conserved sequence and function (Figure 7.2). They have another remarkable property in that they are colinear. Colinearity means that the genes are arranged on the chromosome in the same order as the body regions that they control. In Figure 7.2, they are presented with the genes coding for anterior parts of the body on the left, those coding for central parts in the middle, and those coding for

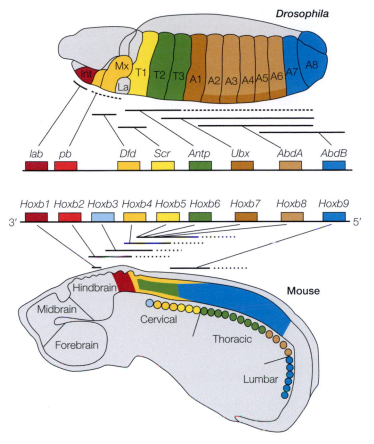

Figure 7.3   Hox genes have preserved colinearity of chromosomal position and function in body plan over hundreds of millions of years. (After Wilkins 2002.)

posterior parts on the right. There is no apparent reason inherent in the logic of gene expression that the Hox genes should be colinear. It may be a pattern inherited from their origin about 600 mya, when colinearity may have had a function it has now lost. It has been preserved throughout most of the animal kingdom; for example, the Hox genes determine similar body regions in flies and mice (Figure 7.3).

As seen in the lower part of Figure 7.2, the entire Hox cluster was duplicated twice in the lineage leading to vertebrates. That happened about 500 mya. It resulted in an expanded toolkit with four copies (*Hoxa*, *Hoxb*, *Hoxc*, *Hoxd*) that could be applied to control new functions without losing control over already-established functions, such as specifying the layout of the body axis. The development of the tetrapod limb (Figure 7.4), which was established about 350 mya in the precursors of the amphibians that colonized land, is controlled by genes in the first and fourth copies in a process that preserves colinearity in the limb, thus reflecting the spatial deployment of genes in the development of the body axis and using combinatorial

Figure 7.4 The evolutionary redeployment of a previously evolved toolkit: colinearity of Hox genes controlling verte-brate limb development. Hox paralogue group, some of the *a*, *b*, *c*, *d* copies of the Hox gene with a given number. (After Gilbert 2013.)

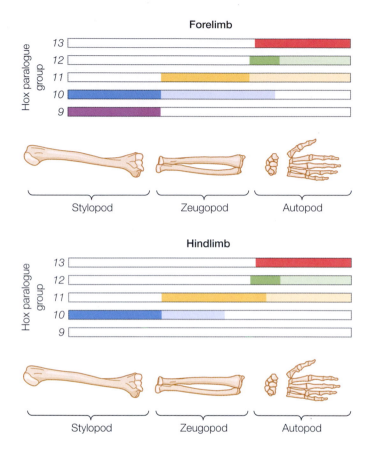

control. If *Hoxd9* and *d10* are expressed, the stylopod (humerus, femur) is formed. If *Hoxd10* and *d11* are expressed, the zeugopod (radius and ulna, tibia and fibula) is formed. Finally, if *Hoxa13*, *d13*, *a12*, *d12*, *d10*, and *d11* are expressed, the autopod (wrist and ankle, fingers and toes) is formed (Gilbert 2013). The sequence proceeds from proximal to distal, mirroring the body axis as it develops from anterior to posterior.

The evolution of the female reproductive system

As with the vertebrate limb, developmental control over the placental female reproductive system involved the redeployment of a previously evolved toolkit to structure a new set of tissues and organs. A comparison of the basic features of female reproductive tracts in amphibians, birds and reptiles, monotremes, marsupials, and placentals makes clear just what had to be accomplished (Figure 7.5) and roughly when it happened in evolutionary history. In the ancestor of the monotremes (>170 mya), eggs evolved reduced yolk, and there were some uterine secretions. In the ancestor of the marsupials (>150 mya), the cloaca was divided, the oviducts differentiated, an eggshell was no longer formed, the placenta originated, and the uterus began to nourish the embryo. In the ancestor of the placentals

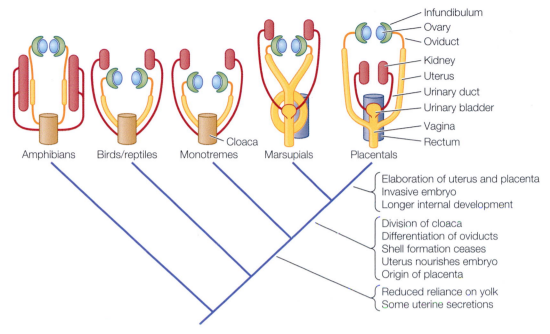

Figure 7.5   The female reproductive system of monotremes and other amniotes is radically different from that of placentals. (After Wagner and Lynch 2005.)

(>105 mya), the uterus and placenta were elaborated, embryonic stem cells evolved the ability to invade the endometrium and take control of some placental function, and the period of internal development—the gestation period—was greatly extended.

Developmental control of these new structures is achieved with the *Hoxa* genes that earlier in development laid down the body axis (Figure 7.6). *Hoxa9* specifies the fallopian tubes, *Hoxa10* specifies the upper uterus, *Hoxa11* specifies the lower uterus, and *Hoxa13* specifies the vagina. *Hoxa10* and *Hoxa11* also aid in endometrial maturation and down-regulate the immune system in the endometrium to help the embryo implant. *Hoxa13* also helps control the development of the umbilical

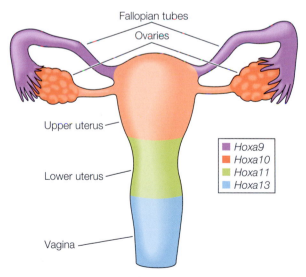

Figure 7.6   The placental female reproductive tract is controlled by the same genes that lay down the body axis earlier in development. (After Lynch and Wagner 2005.)

arteries. As with the vertebrate limb, the same toolkit with its complex switches and combinatorial control is deployed here in a new context in which colinearity is preserved.

One mechanism involved in the innovations: Transposon insertion

How a control gene acquires jurisdiction over a new developmental district is a puzzle that in this case is partially solved (Wagner and Lynch 2010). Transposons are mobile genetic elements that can change position in the genome. When a transposon with a binding site for a transcription factor inserts into the control region upstream from a target gene, that gene comes under control of the transcription factor. In endometrial stromal cells, the transcription of prolactin is controlled by an enhancer derived from a transposable element (MER20) that invaded the mammalian genome about 170 mya when placentation evolved. Such movement of the entire control sequence is much easier to envision than the de novo construction of a new control sequence through a series of mutations.

In all, about 1500 genes have been recruited into endometrial expression in placental mammals, and 13% of them—about 200—are close enough to the MER20 insertion to come under its control. That region binds transcription factors essential for pregnancy, regulating gene expression in response to progesterone and cyclic AMP (Lynch et al. 2011).

That is not the only case in which the insertion of a transposon aided a major evolutionary innovation. It also happened at the origin of the vertebrate adaptive immune system, enabling the somatic recombination that generates the diversity of cell lineages needed for clonal selection and immune memory (see Chapter 4).

■ SUMMARY

Placental reproduction involves a set of evolutionary innovations that required the rewiring of a large genetic network. The placental female reproductive tract develops under the control of the same genes that lay down the body axis earlier in development. Some of the innovation was made possible by the insertion of a transposable element containing binding sites for transcription factors.

The evolution of invasive placentas

The implantation of the embryo into the endometrium to initiate the development of the placenta is an aggressive process. Embryonic tissue invades maternal tissue, giving more invasive embryos a fitness advantage early in life through increased control of nutrition and growth. Placental invasion is ancient; it appears to be the ancestral condition inherited from the origin of placental mammals. Interestingly, however, the degree of invasiveness now varies among major groups, having been lost to a greater or lesser degree in some. One reason for such loss may be the cost of placental invasion. Two such costs are increased risks of pre-eclampsia and metastatic cancer. For the former, greater embryonic control over maternal blood pressure increases the risk of dangerously high blood pressure (pre-eclampsia). For the latter, every cell in the body retains the genetic information needed to direct stem cell movements early in life, and the properties of invasive stem cells are very similar to those of metastatic cancer cells.

During implantation, the outer cells of the pre-embryo—the trophoblast—bore into the endometrium by secreting enzymes that dissolve the tissue protecting the maternal blood vessels. The endometrium responds by both limiting and enabling this invasion. The trophoblast then becomes part of the placenta, where it secretes hormones that mimic maternal hormones. Those hormones influence the level of maternal blood sugar and the dilation of the maternal arteries in the placenta, thus helping to regulate the delivery of nutrients. The degree of invasion and the consequent ability of embryonic tissue to control nutrient flow both depend on placental morphology, which varies greatly among the major groups of placental mammals.

Placental morphology

The critical features of placental morphology are those that determine how many layers of cells and tissue separate fetal blood from maternal blood and how intimately fetal tissue and maternal tissue are intertwined over how large a surface area. Prior to formation of the placenta, six layers of tissue separate maternal and fetal blood (Figure 7.7). During placentation, one, two, or three of the maternal layers can be lost. The fetus contributes all three of its extra-embryonic membranes to form the placenta: the endothelium lining the allantoic capillaries; connective tissue in the chorioallantoic mesoderm; and, from the trophoblast—the outermost layer of the placenta—the chorionic epithelium. The tissue layers contributed by the

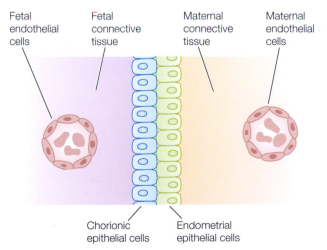

Figure 7.7   Six cells layers, three each from fetus and mother, can potentially contribute to the formation of the placenta. (After Bowen 2011.)

(A) Epitheliochorial (cows, pigs, horses)

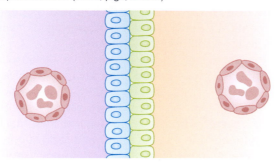

(B) Endotheliochorial (dogs, cats)

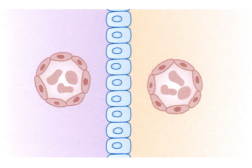

(C) Hemochorial (humans, rodents)

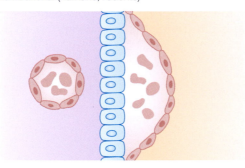

Figure 7.8   Placental classification based on cell layers. This differentiation is what a veterinarian encounters when treating a variety of pregnant species. (After Bowen 2011.)

mother can include the endothelium lining the endometrial blood vessels, connective tissue in the endometrium, and the endometrial epithelial cells.

Depending on the number of maternal layers that are retained, the type of placenta that results will be epitheliochorial, endotheliochorial, or hemochorial (Figure 7.8). In cows, pigs, and horses, whose placentas are epitheliochorial, maternal and fetal blood remains separated by 6 layers

(A) Diffuse

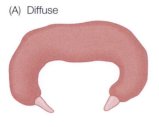

Figure 7.9 Both the shape and the number of tissue layers contribute to the intimacy of contact between maternal and fetal blood. (A) Diffuse, horses and pigs; (B) cotyledonary, cattle and other ruminants; (C) zonary, dogs, cats, seals, bears, and elephants; (D) discoid, primates and rodents. Light pink, amniotic sac; dark pink, placenta. (After Bowen 2011.)

(B) Cotyledonary

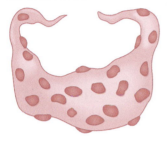

(C) Zonary

(D) Discoid

of tissue. In dogs and cats, whose placentas are endotheliochorial, the intervening tissue is reduced to 5 layers. In humans and rodents, whose placentas are hemochorial, the growing chorionic villi erode through the maternal epithelium, forming a hemochorial placenta in which the fetal chorionic epithelium is bathed directly in maternal blood. Here the intervening tissue has been reduced to just 3 layers. The resulting placenta can have one of four shapes (Figure 7.9): diffuse, occupying most of the surface of the embryonic sac; cotyledonary, with multiple, discrete attachments; zonary, forming a band around the fetus; or discoid, attaching at one point and smaller than the fetus. Humans, apes, monkeys, and rodents have discoid, hemochorial placentas that contrast strikingly with the types of placentas found in other mammals.

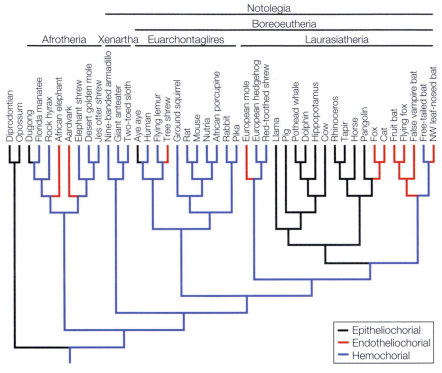

Figure 7.10 The evolutionary history of tissue layers in placentas. (After Wildman et al. 2006.)

The history of placental layers and shapes

The number of cells layers separating maternal and fetal blood in the human hemochorial placenta is an ancient, primitive state shared with manatees, elephant shrews, armadillos, and rodents (Figure 7.10) but not with horses, cows, cats, and bats. A study of 109 species from all mammal orders revealed that the fewer the number of tissue layers separating maternal and fetal blood, the shorter the length of gestation for a given maternal body mass (Capellini, Venditti, and Barton 2011). The shift from noninvasive, epitheliochorial placentas to intermediately invasive endotheliochorial placentas, to highly invasive hemochorial placentas reduced gestation time on average by about 35 days. Variation in placental layers had, however, little effect on the relative brain size of infants or adults.

The external morphology of the human placenta—its discoid shape—is also ancient and primitive, but its internal villi—the tree-like extensions of the fetal circulatory tract that are bathed in maternal blood—are derived, shared with primates, whales, and horses, but not with bats, rodents, and elephants (Figure 7.11). In tuning the intimacy of the fetal-maternal connection, evolution has clearly had several major traits with which to work, resulting in different degrees of fetal and maternal opportunities for control in different major groups.

(A) History of placental shape

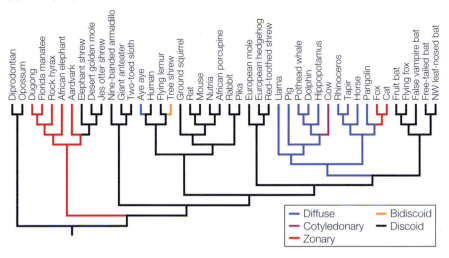

(B) History of internal morphology

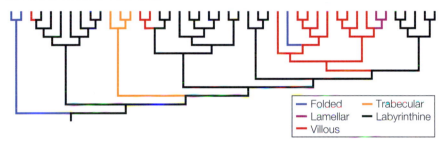

Figure 7.11   The evolutionary history of (A) placental shape and (B) internal morphology. (After Wildman et al. 2006.)

Placental morphology and risk of pre-eclampsia

Humans share with the great apes especially deep extravillous invasion and remodeling of the spiral arteries by cells from the embryo (Table 7.1). It is striking that great apes also share with humans the risk of pre-eclampsia, a condition not observed in species with slightly less invasive placentas, such as rodents and armadillos.

Genomic analysis of the genes involved in risk of pre-eclampsia and trophoblast invasion shows that both classes of genes experienced positive selection at the origin of the Hominidae (the great apes and humans) and the Homininae (the smaller group containing humans, chimpanzees, and gorillas). Of 18,000 genes examined for signals of selection, 295 had such signals in the branch leading to the Hominidae, and 264 had them in the branch leading to the Homininae. These sets were enriched for genes that function in trophoblast invasion and risk of pre-eclampsia (Crosley et al. 2013).

TABLE 7.1   Differences in human placentation compared with nonhuman primates and other placental mammals

Human	Other primates	All placental mammals
Primary interstitial implantation	Occurs only in great apes.	Occurs in guinea pigs and some other rodents. Eccentric implantation in which the uterus forms an invagination to surround the blastocyst is more common (e.g., mouse).
Yolk sac placenta absent at all developmental stages	A transient yolk sac placenta occurs in strepsirrhines. It is absent in all haplorhines.	Most mammals have at least a transient yolk sac placenta, and many (e.g., mouse) retain a yolk sac placenta to term.
Allantoic sac replaced by an allantoic stalk	Considered to be a defining feature of haplorhines, but some New World monkeys have a rudimentary sac. A large allantoic vesicle occurs in strepsirrhines.	The size of the allantoic vesicle varies greatly, and it is altogether absent in some mammals (e.g., mouse).
Villous placenta with an intervillous space	Occurs in Old World monkeys and great apes. Tarsiers have a labyrinth. The trabecular placenta of New World monkeys is intermediate between the two.	Unusual outside the primates, but does occur in armadillos and anteaters.
Interhaemal barrier haemomonochorial (i.e., with a single layer of syncytiotrophoblast at term)	Shared by all haplorhine primates.	Not uncommon (e.g., guinea pig). Some species (e.g., rabbit, mouse) have 2–3 layers of trophoblast.
Two routes of trophoblast invasion	Invasion by the interstitial route shared only by great apes.	Highly variable.
Deep trophoblast invasion reaching inner third of myometrium	Shared only by great apes.	Conditions vary with even deeper invasion of vessels known (e.g., some bats, guinea pig).

Source: After Carter and Pijnenborg 2011.

Placental invasiveness and risk of metastasis

There are striking similarities between the embryonic cells that invade the placenta and metastatic cancer cells that can move through the body to invade tissues and organs. The similar cellular mechanisms include particular cell adhesion molecules, certain features of the extracellular matrix, and expression of metalloproteinases that can degrade the extracellular matrix to enable migration. The cells that help build the placenta also resemble metastatic cancer cells in having the ability to stimulate angiogenesis. These observations suggest that understanding the maternal mechanisms that have evolved to control trophoblast invasion may be

TABLE 7.2 Proto-oncogenes that are expressed by both normal trophoblast and malignant nontrophoblastic cells

Proto-oncogenes	Expression	
	Normal trophoblasts	Malignant tumors
c-ras (RAS)	Villous cytotrophoblasts	Colorectal
c-kit (KIT)	Villous cytotrophoblasts	Breast
		Testis
c-jun (JUN)	Cytotrophoblasts, extravillous cytotrophoblasts	Lung
		Kidney
c-met (MET)	Cytotrophoblasts	Colon
		Gastric
c-fos (FOS)	Villous cytotrophoblasts	Cervical
	Extravillous cytotrophoblasts	Breast
c-myc (MYC)	Cytotrophoblasts, extravillous cytotrophoblasts	Breast
		Endometrium
c-erb-B1 (ERBB1, HER1)	Cytotrophoblasts (before 6 weeks)	Breast
	Syncytiotrophoblasts (after 6 weeks)	Ovary
c-erb-B2 (ERBB2, HER2)	Extravillous cytotrophoblasts	Breast
		Gastric
		Cervical
		Ovary
c-sis (SIS, PDGFB)	Extravillous cytotrophoblasts	Leukemia
		Osteosarcoma
c-fms (CSFIR)	Extravillous cytotrophoblasts	Liver
	Syncytiotrophoblasts	
c-Abl (ABLI)	Trophoblasts	Leukemia
		Ovarian

Source: After Ferretti et al. 2007.

one way to get important insights into methods for controlling metastases. The list of genes that are expressed both in trophoblast and in metastases is impressively long and continues to grow (Table 7.2).

Invasive placentation is primitive in placental mammals: it was present in the common ancestor. However, it has been suppressed by secondary evolution in several mammalian lineages in which there should be mechanisms that also resist cancer cell invasion. In fact, animals with less invasive placentation—the cows and horses—have less risk of three types of metastatic cancer than do animals with more invasive placentation—the dogs and cats—as shown in Figure 7.12.

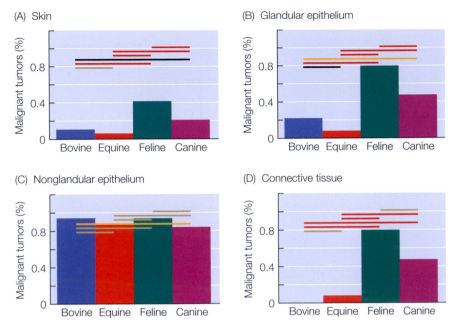

Figure 7.12   Percentage of each type of malignant tumor in four types of tissue in four types of animal. The bars indicate significant differences: tan, nonsignificant; black, $p < 0.05$; orange, $p < 0.005$; red, $p < 0.001$. Nonglandular epithelial cancers (C) have uniformly high malignancy rates. For the other three (A, B, and D), however, the hoofed animals with noninvasive placentation have significantly lower rates than do the carnivores, which have invasive placentation. (After D'Souza and Wagner 2014.)

■ SUMMARY

Mammalian placentas vary widely in the number of tissues layers separating fetal and maternal blood, in the shape of the structures that mediate exchange, and in the degree of fetal invasion of maternal tissue. Human placentas are a mix of ancient and derived features. They share especially deep invasion with chimpanzees and gorillas. That derived feature appears to be associated with risk of pre-eclampsia. Although invasive placentas clearly promote fetal growth, they also carry another risk: they predispose cells to be able to metastasize in cancer progression.

Parent-offspring conflict, genomic imprinting, and pregnancy

The maternal-fetal relationship has traditionally been seen as completely positive, designed by evolution for effective parental care of a grateful and cooperative recipient offspring. That is not the picture painted by Williams

(1957) and Hamilton (1964a,b), whose insights focus on selection operating on genes at the potential expense of individual survival to explain the evolution both of aging and of cooperation and conflict among relatives. In the case of conflicts among relatives, the chain of ideas is long, from Hamilton's concept of kin selection to Trivers's (1974) and Haig's (1993) thoughts on how parent-offspring conflict helps explain pre-eclampsia and gestational diabetes to Haig's (e.g., Haig 2000) insight that disturbances of parent-of-origin imprinting affect fetal growth, gestation length, size at birth, and perhaps some crying at night (Haig 2014). These views fundamentally change how we conceive of the parent-offspring relationship: its normal state is an equilibrium in an evolutionary tug-of-war, disturbances to which elicit pathologies. We now trace these ideas from kin selection through parent-offspring conflict to their application to some diseases of pregnancy.

The basic idea of kin selection

If what matters to evolution is not individual survival but the increase in frequency of genes, it will pay the genes in an individual to risk even death if more copies of themselves get into the next generation because of that decision than if the individual did not so act. Note that the costs and benefits of an act are to be weighed against each other in genetic currency and that genes exist not only in an individual but also in its relatives. It follows that if an individual can help its relatives survive and reproduce and if the increase in the number of its genes in the next generation through offspring of relatives is greater than the decrease in the number of genes transmitted through its own offspring, helping behavior will be selected.

That idea can be formalized in an inequality that states the condition for acts of helping to evolve. Let B represent the increase in the recipient's fitness as a result of an act; C represent the decrease in the donor's fitness as a result of the act; and r represent the coefficient of relationship, the probability that a gene in the donor and a gene in the recipient are identical by descent from a common ancestor. For example, in a diploid sexual population like our own, $r = 0.5$ for full siblings and 0.25 for half siblings. The condition for helping is then $B/C > 1/r$, or $B \times r > C \times 1$. In other words, the increase in the relative's fitness as a result of the act multiplied by the relationship to the relative ($B \times r$) must be greater than the decrease in the donor's fitness multiplied by its relationship to itself ($C \times 1$).

Traits that have been proposed as instances of kin-selected cooperation include the alarm calls of birds, ground squirrels, and primates; the guarding behavior and vigilance of meerkats; helping at the nest in birds such as acorn woodpeckers, Florida scrub jays, and pied kingfishers; and suppressed reproduction in highly social species such as bees, ants, wasps, meerkats, dwarf mongooses, hyenas, and naked mole rats. There are alternative explanations for some of these traits, including ecological constraints that force individuals to make the best of a bad lot by choosing the least bad of several options.

The basic idea of parent-offspring conflict

The theory of parent-offspring conflict builds directly on the idea of kin selection. Whereas a parent in a diploid, sexually reproducing species is only 50% related to each of its offspring, an offspring is 100% related to its self, 50% related to full siblings, and 25% related to half siblings. An offspring should therefore seek to extract more investment from its parents than its parents have been selected to give, trying to force the parents' investment upward until the offspring's inclusive fitness—its direct fitness plus the contributions to its fitness that come through current or future siblings—is maximized.

Parent and offspring are expected to disagree over how long the period of parental investment should last (both gestation and suckling in mammals); over the amount of parental investment that should be given; and over the altruistic or selfish behavior of offspring toward relatives, siblings in particular. Such conflict is expected to increase during the period of parental care, culminating in weaning conflicts, and offspring are expected to deploy both biological and psychological weapons in conflict with their parents.

Parent-offspring conflicts in pregnancy

The idea that offspring can be in conflict with their parents sheds new light on placental invasion and the diseases of pregnancy. The fetus is expected to try to extract more from the mother than the mother has been selected to give, and fetal tissue in the placenta can manipulate maternal physiology via hormone production. Such manipulations are made easier when fetal tissue invades maternal tissue in the placenta and when the maternal blood supply is directly exposed to fetal tissue without any intervening layers of maternal tissue that might function to control the interaction. The mother is expected to resist fetal manipulation that goes beyond her own interests, with the usual result being an equilibrium in which both mother and offspring are healthy.

There are two main paths to increasing fetal provisioning. The fetus can act to increase blood flow by expanding spiral arteries and increasing maternal blood pressure. If that goes too far, the result can be dangerously high maternal blood pressure—pre-eclampsia—which can be life-threatening. The fetus can also act to increase the concentration of sugar in maternal blood by secreting placental hormones that cause insulin resistance in maternal tissues. If that goes too far, the result can be gestational diabetes. Neither pre-eclampsia nor gestational diabetes is selected. Both conditions are bad for the mother and bad for the infant. They are seen here as by-products caused by disruptions of an underlying conflict. One of the mechanisms that can mediate such conflicts is genomic imprinting.

Evolutionary conflicts and genomic imprinting

Gene expression can be silenced by DNA methylation of regulatory regions (enhancers and promoters), establishing repressive heterochromatin that can be maintained through cell divisions. The resulting stable repression of gene expression, known as gene imprinting, can affect either one or both alleles of the gene. A particularly interesting case of imprinting, called parent-of-origin imprinting or genomic imprinting, occurs when maternal and paternal alleles are processed differently. In this case, alleles imprinted in the germline are turned off: they are not expressed in the fetus during pregnancy or in the childhood development of the offspring. When the child matures into either a male or a female, the imprinting pattern in the germline is adjusted to represent the interests of that sex.

Hamilton's idea of kin selection also informs our understanding of parent-of-origin imprinting. The mother is 50% related to each of her offspring, but if she has offspring with other males, only this offspring is 50% related to the current father, and future offspring by other males will not be related to him at all. Thus, to the degree that mating is polygamous, paternal genes will be selected to extract more from the mother than she is selected to give, and maternal genes are selected to resist. The father turns off genes that down-regulate growth; thus, transcripts of genes inherited from the father enhance growth. The mother turns off genes that up-regulate growth; thus, transcripts of genes inherited from the mother inhibit growth. The normal state is an equilibrium in which both mother and offspring are in reasonably good condition.

The conflict is revealed when the action taken by one parent is canceled, which can be done by disrupting the imprinting. In mice, such disruptions can be made by genetic manipulations. Disruptions of the paternal imprinting pattern, which allow maternal interests to be expressed unimpeded, decrease weight at birth by 10%. Disruptions of the maternal imprinting pattern, which allow paternal interests to be expressed unimpeded, increase weight at birth by 10%. Similar patterns are seen in humans when parent-of-origin imprinting patterns are disrupted by rare mutations in one parent but not the other and when copies of genes coming from one parent are duplicated or deleted. This interpretation is now well supported for the genes for insulin-like growth factor (*igf2*) and its receptor (*igf2R*) and for two other genes (*CDKNIC* and *GRB10*), but not yet for other genes imprinted by the sex of origin (Haig 2004).

Evolutionary conflicts and mental disorders

Although many of the genes that show parent-of-origin imprinting are expressed in the fetus and in the fetal tissue of the placenta, others are expressed in the brain, where they are thought to continue to mediate parent-offspring conflict via behavior. We discuss these effects in Chapter 9 because their disruption appears to contribute to the mental disorders of autism and schizophrenia.

■ SUMMARY

Natural selection acts to increase the frequency of genes, not the survival of individuals. Genes can influence their fate by shaping actions that affect the reproductive success of relatives. Parent and offspring are in conflict over reproductive investment: the offspring is selected

TABLE 7.3 The major hypotheses for the evolution of menstruation

Hypothesis	Claim	Comments
Sperm-borne pathogen removal	Menstruation occurs to protect the uterus against colonization by pathogens transported by sperm.	Menstruation occurs weeks after copulation; the problem of sperm-borne pathogens is not unique to menstruating species.
Energy conservation	Menstruation/endometrial resorption is energetically less costly than maintaining a differentiated endometrium.	Maintaining a differentiated endometrium is not the alternative in other species. It also would not allow for ovulation, sperm transport, or sperm capacitation.
Nonadaptive consequence of spontaneous decidualization	Menstruation occurs as a nonadaptive consequence of spontaneous decidualization of the endometrium.	This is consistent with known consequences of decidualization and is the position adopted here. The remaining problem is to explain the evolution of spontaneous decidualization.
Uterine preconditioning	Menstrual "preconditioning" protects uterine tissues from the hyperinflammation and oxidative stress associated with deep placentation in humans.	Claim ignores why menstruation may have evolved in ancestral primate species and menstruating nonprimates; claim ignores benefits that spontaneous decidualization itself might provide; there is no experimental evidence that endometrial shedding confers protection against oxidative stress during pregnancy.

Source: After Emera, Romero, and Wagner 2011.

to demand more than the parent is selected to give. Maternally and paternally derived genes are in similar conflict within the genome of an offspring, which appears to explain the existence of parent-of-origin (genomic) imprinting. In both cases, the normal condition is an equilibrium established by an evolutionary tug-of-war. Disruption of that equilibrium can manifest as unselected, by-product pathology.

Menstruation

Menstruation is often uncomfortable, is sometimes painful, and can be socially awkward. Why did it evolve? Answers to that type of question see the trait as either directly selected or as the by-product of another process (Table 7.3). It has been suggested that menstruation evolved to eliminate parasites, to save energy, as a nonadaptive consequence of spontaneous decidualization of the endometrium, or to protect the uterus from inflammation. There are problems with the three adaptive hypotheses (Emera, Romero, and Wagner 2011). All mammals encounter pathogens in their reproductive tracts, but only a few of them menstruate. Eliminating pathogens is probably not the reason for menstruation. Neither is saving energy: the energetic costs of maintaining a differentiated endometrium are not the alternative in nonmenstruating species, whose placentation is often triggered by mating.

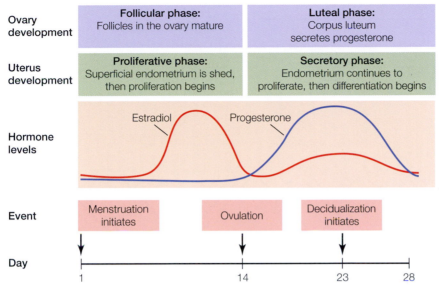

Figure 7.13 The human menstrual cycle. (After Emera, Romero, and Wagner 2011.)

Before analyzing the evolutionary history and possible reasons for the evolution of menstruation, it will help to review how hormones coordinate events in the ovary and uterus (Figure 7.13). In the first half of the menstrual cycle, follicles in the ovary mature, the superficial endometrium of the uterus is shed in menstruation, and then estradiol levels rise and proliferation of the endometrium begins. Ovulation marks the middle of the cycle. In the second half of the cycle, the corpus luteum in the ovary secretes progesterone; in the uterus, the endometrium continues to proliferate, and decidualization, stimulated by progesterone, begins, preparing the endometrium to receive an implanting embryo. If an embryo does not implant, progesterone levels fall, the differentiated endometrial stromal cells undergo apoptosis, and menstruation starts.

The comparative biology and evolutionary history of menstruation

Menstruation has evolved three times: once in primates, once in elephant shrews, and at least once in bats (Figure 7.14). Comparing the reproductive biology of the species that menstruate reveals several shared features that offer clues to its function. They all have spontaneous decidualization; invasive, hemochorial placentation; extended mating periods; spontaneous ovulation (not induced by mating); and one or two well-developed offspring per pregnancy.

In mammals that do not menstruate, decidualization occurs when the embryo implants. In mammals that do menstruate, decidualization is

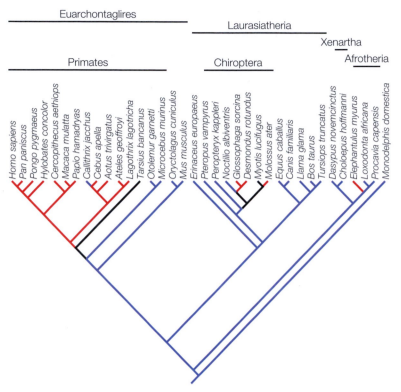

Figure 7.14 The phylogenetic distribution of menstruation in primates, bats, and elephant shrews. Pink, menstruating species; blue, non-menstruating species; black, menstruation state not yet determined. (After Emera, Romero, and Wagner 2011.)

spontaneous: it is under maternal control and occurs whether an embryo implants or not. Thus, to understand the evolution of menstruation, we need to understand why spontaneous decidualization evolved. At least two hypotheses can explain it, and they are not mutually exclusive:

1. Spontaneous decidualization is a maternal response to protect the mother from the invasive embryo.
2. Spontaneous decidualization is a mechanism that allows the mother to discard defective embryos.

At this point, we do not know whether either of these hypotheses is correct, but we do have other observations that support the second one without refuting the first. It is clear that the female reproductive tract contains mechanisms that control the quality of gametes and concepti. Oocytes form in the developing ovaries of 3-month-old fetuses, and they do not form thereafter. Oocytic atresia (the destruction of oocytes through apoptosis) starts in that third month of pregnancy, when there are about 7 million oocytes in the ovaries. By birth, about 6 million oocytes have been discarded; by menarche, only thousands are left; and at menopause, the number is near zero. Atresia appears to target both nuclear and mitochondrial mutations. This process of screening out potentially defective progeny continues after conception. The proportion of pregnancies that end in early spontaneous abortions (miscarriages) is estimated to range from 30% to 75%. Of pregnancies that last long enough to be clinically recognized, 10%–20% miscarry: most of those have chromosomal abnormalities. The widespread use of ultrasound has revealed that about 70% of gestations that begin as twins end as singletons, one twin having died and been reabsorbed early in pregnancy. These processes serve to eliminate defective offspring and reduce the reproductive costs of twins, allowing the mother to concentrate her reproductive investment in a few high-quality, healthy offspring with good chances of surviving to maturity.

■ SUMMARY

Menstruation probably did not evolve to eliminate parasites or save energy. It does appear to be a by-product of spontaneous decidualization of the endometrium, which occurs under maternal control whether or not she has mated and, if she has, whether or not an embryo implants. It may function to protect the mother from invasive embryos, and it may be a mechanism that allows the mother to discard defective embryos. Some evidence supports the second possibility without eliminating the first.

Menopause

Menopause is puzzling from an evolutionary perspective. If reproductive success is the measure of fitness, why should an organism stop reproducing? Menopause is rare in mammals, and it is a derived condition in humans that originated after we shared ancestors with chimpanzees, bonobos, and gorillas because they do not experience menopause. Two cetaceans—pilot whales and killer whales—do have menopause and, like us, close, unusually long maternal-offspring interactions.

Menopause results from an evolved change in survival, not in the fertility schedule (Figure 7.15). Both human hunter-gatherer (!Kung and Aché) and chimpanzee females can reproduce to about age 50, but chimpanzee females do not experience menopause and do not survive past age 55, whereas human females do experience menopause at 45–55 and survive at least another 20 years.

There are at least four hypotheses for the evolution of menopause—the mother, the grandmother, the by-product, and the self-domestication hypotheses—and they are not mutually exclusive. Neither are any of them as yet soundly established, although the available evidence suggests that they may have functioned in combination.

The mother hypothesis claims that menopause evolved as part of terminal reproductive investment. As mothers age, they encounter an increasing risk of dying in childbirth. Avoiding that risk, although costing an offspring, would ensure that the mother would survive for the period of parental care

(A) Chimpanzees

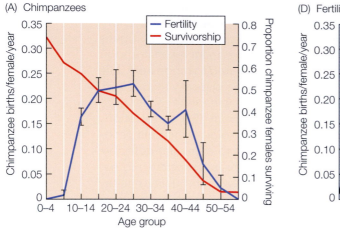

(D) Fertility comparison

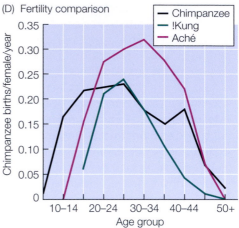

(B) !Kung

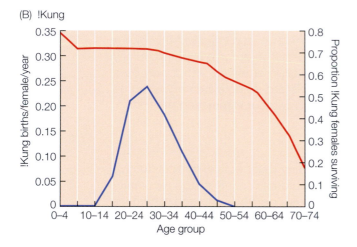

(C) Aché

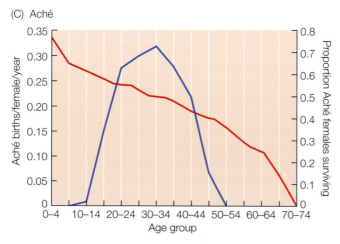

Figure 7.15 Menopause results from an evolved change in survival, not fertility. The fertility curves of hunter-gatherers overlap with those of chimpanzees rather closely (D), but human hunter-gatherers (B and C) survive much longer after having their last child than do chimpanzees (A). (After Thompson et al. 2007.)

needed for the last offspring (Williams 1957). Although this hypothesis makes some sense, it is also clear that females survive longer than the period during which their parental care would be necessary.

The grandmother hypothesis claims that menopause evolved to reduce the increasingly compromised reproduction of aging grandmothers so that they could gain more fitness by helping their offspring raise their grandchildren (Hawkes et al. 1998). Lee (2003b) has shown that in theory such intergenerational transfers from the old to the young can support selection to extend postreproductive survival. The evidence on grandmother effects is mixed, however, with some studies finding that the presence of grandmothers improves and others finding that it is either neutral or detrimental to the survival of grandchildren (Lahdenperä et al. 2004; Shanley et al. 2007; Madrigal and Meléndez-Obando 2008; Lahdenperä et al. 2011; Lahdenperä et al. 2012).

The by-product hypothesis claims that menopause evolved as a by-product of quality control of oocytes, reasoning that having high-quality offspring early in life would outweigh having a few more offspring late in life, especially if the aging of the quality control filters means that offspring born late in life are more likely to be defective (Stearns and Ebert 2001). Not enough is known at this point about the efficacy of quality control and its pattern of aging to evaluate this hypothesis quantitatively.

The self-domestication hypothesis claims that postreproductive life resulted from improvements in survival caused first by social behavior and then extended by the agricultural revolution. The existence of an extended period of postreproductive life then created the opportunity for natural selection to modify the behavior of postreproductive individuals to enhance their fitness by contributing to the survival of grandchildren. This view is compatible with all the data supporting the grandmother hypothesis, but it sees causality as working in the other direction: first, postreproductive life as a by-product of culture, followed by selection for grandparental behavior (Ellison and Ottinger 2014).

■ SUMMARY

Menopause is a recent evolutionary innovation in the lineage leading to humans that may have evolved for a combination of reasons. Included are the terminal reproductive investment of mothers, the help that nonreproductive grandmothers can give to their offspring, as a by-product of quality control of gametes, and as a by-product of cultural evolution that improved survival to the point where a significant proportion of individuals experienced postreproductive life.

Upright posture and childbirth

Human childbirth is especially difficult because the pelvis has been modified for bipedalism. The difficulties of childbirth cause significant maternal

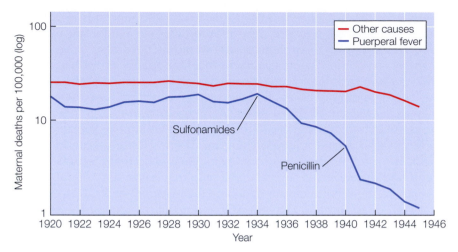

Figure 7.16 Childbirth has historically been very risky for mothers, as shown in the number of maternal deaths (log scale) per 100,000 births in England and Wales from 1920 to 1945. (After Loudon 2000.)

mortality, which has only recently declined in developed countries and remains high in undeveloped countries. If human infants developed to the same stage at birth as is attained by our closest relatives, they would have longer gestations and be larger at birth. However, the female pelvis has constrained the size of the infant cranium that can pass through the birth canal and thus the size and developmental stage of the brain at birth. Whereas the infant cranium can be delivered directly in chimpanzees, it must rotate to pass through the birth canal in humans. As a result, most human infants are born facing backward; in other primates, they are born facing forward, a position in which the mother finds it much easier to catch and support the newborn. This difference may imply that human females have had assistance in childbirth since roughly the time that bipedalism evolved.

Childbirth is risky

Even after Ignaz Semmelweiss discovered the importance of sanitation and sterile conditions during deliveries in the obstetric wards of nineteenth-century Austria, maternal mortality in childbirth remained high, and it did not start to decline until Jacob Lister introduced carbolic acid as an antiseptic in the 1860s. Even then, it remained at 300–700 deaths per 100,000 births in developed countries and did not decline further until the introduction of antibiotics in the 1930s and 1940s removed most of the threat posed by puerperal fever (Figure 7.16). In undeveloped countries, particularly in Africa and South Asia, as many as 1% of women giving birth continue to die as a result.

(A) Chimpanzee

(B) Human

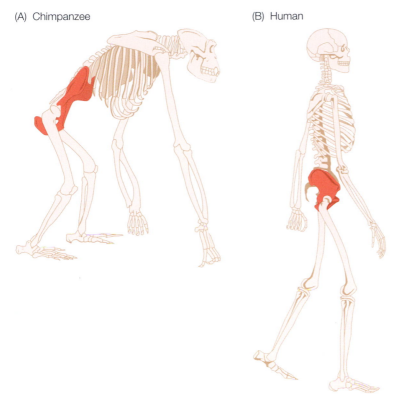

Figure 7.17   The human pelvis was fundamentally remodeled to support bipedalism and rapid long-distance pursuit of prey. (© roadrunner/Fotolia.)

Human bipedalism and brain size combine to make birth difficult

The remodeling of the pelvis that occurred when bipedalism evolved was dramatic (Figure 7.17). The remodeled pelvis supported both walking and the rapid long-distance running in pursuit of prey that is also thought to explain the evolution of our hairlessness, dark skin color, and increased ability to regulate body temperature by sweating. Bipedalism evolved 4–7 mya in *Australopithicus* or earlier; there followed a considerable delay before the narrow birth canal evolved in *Homo habilis* about 2–3 mya. During that period, the brains of our ancestors increased considerably in size.

The infant cranium could not be any larger and still fit into the birth canal (see Figure 1.7); to transit the birth canal, the infant's head has probably had to rotate since bipedalism originated in *Australopithecus* or earlier. As mentioned above, the rotation causes humans to be born facing backward, unlike most primates, where the mother can catch her infant and rapidly clean its nose to help it breathe. A human mother has a harder time doing that, and both mother and infant benefit from the assistance at the birth of someone in the role of a midwife (Rosenberg and Trevathan 2002).

Adding to the morphological and evolutionary tensions that characterize human birth is that human brains, despite pushing the envelope of the birth canal, are small at birth relative to their adult size (Harvey and Clutton-Brock 1985). They grow after birth more than do the brains of any other primate infants. That period of brain growth, which lasts until about age 7, defines the developmental period that we call childhood. It does not occur in the great apes, in which infants transition much more directly into adolescence.

■ SUMMARY

Upright posture, which evolved 4–7 mya, and rapid long-distance pursuit of prey, which followed about 2–3 mya, caused the remodeling of the pelvis. The consequent narrowing of the birth canal has constrained size at birth by directly limiting the diameter of the infant cranium that can pass through it. Selection to increase infant size at birth is balanced by selection acting through maternal and infant mortality in childbirth: there are trade-offs between the health and development of the infant and the survival of the infant and mother. These morphological constraints have contributed to the evolution of a uniquely long human childhood, during which time the brain grows to adult size.

Mismatch

When natural selection acts on a population for a long time in a stable environment, it produces near-optimal adaptations to the stable environmental factors, which include sources of food, pathogens, predators, and temperature. Even under these ideal circumstances, the adaptive solutions are not necessarily the best ones possible because they are constrained by historical contingencies and the range of genetic variants available. They do, however, usually represent the best available matches of phenotypes to the environment, and often the mutations that occur in such circumstances are detrimental and selected against. If the environment then changes, some aspects of the phenotype, and some of the genetic variation that underlies the phenotype, will be "mismatched" to one or more environmental factors. Depending on the degree and nature of the mismatch, the organisms in the population will suffer mild to severe loss of fitness, even death. In humans, we refer to such phenotypes as diseases. Mismatch thus describes both a poor fit of phenotype to environment and a situation in which selection is acting to improve that fit.

Environmental change in humans

In most natural populations, environmental change is slow and often undirected and unpredictable. In modern humans, however, the changes are extremely fast, direct, and predictable in the short term.

The importance of relative rates

Most of the mismatch that causes health problems in modern societies is the result of our culture changing much more rapidly than biology can keep pace. For example, lactose tolerance, a relatively genetically simple trait under reasonably strong selection, has not yet risen to high frequency in many human populations since it began to increase in milk-drinking cultures with the domestication of dairy animals 5000–10,000 years ago (more on this topic below). The rate of biological change of many human traits is quite slow when contrasted with the cultural change of our recent past. We do continue to evolve biologically, but it takes several hundred generations and several thousand years to produce significant change in a trait unless selection is unusually strong.

The many consequences of human niche construction

Many organisms alter their environment to better suit their needs. Familiar examples of this process, known as niche construction, include beaver dams, termite mounds, bird nests, and subterranean burrows. Most of the changes in the modern human environment are either the direct product of niche construction (e.g., urbanization, agriculture, the medical system) or its indirect consequences (e.g., air and water pollution, climate change, extinctions).

We have been constructing our niche for a very long time, but most of the striking features of our current niche were constructed in the relatively recent past, too recently for our biology to have adapted very much. Those of us who live in the postindustrial part of the world experience much less exposure to parasites, pathogens, and symbionts than in the past; a great deal of artificial light that has altered our activity rhythms; profound changes in diet; a much more sedentary lifestyle; and, in many parts of the world, temperature-controlled housing. We now encounter much more sugar, starch, salt, and addictive drugs, including alcohol and tobacco, than did our ancestors. Many of these things produce mismatches between biology and culture, with consequences that can be controlled or compensated, when understood, with changes in technology, behavior, and lifestyle. Because agriculture, industrialization, and postindustrial globalization have caused so many changes in our environments, our reactions to the changes can have many causes, and claims of simple causation should be viewed skeptically. Most of the "diseases of civilization" have complex causes with strong genotype × environment interaction effects. In such cases, simple answers can be misleading.

Dysevolution and the mismatch diseases

When we pass on the cultural factors that cause mismatch diseases, the cultural factors can become more prevalent and more intense; Lieberman (2013) calls this harmful form of cultural change over time *dysevolution*. Mismatch (the initial cause) and dysevolution (its cultural continuation and exacerbation) are thought to be responsible for about 50 noninfectious diseases and other health issues (Table 8.1).

TABLE 8.1   Conditions thought to be caused by mismatch

Acid reflux/chronic heartburn	Depression	Inflammatory bowel disease
Acne	Diabetes (type 1)	Lactose intolerance
Alzheimer's disease	Diaper rash	Lower back pain
Anxiety	Eating disorders	Malocclusion
Apnea	Emphysema	Metabolic syndrome
Asthma	Endometriosis	Multiple sclerosis
Athlete's foot	Fatty liver syndrome	Myopia
Attention deficit hyperactivity disorder	Fibromyalgia	Obsessive-compulsive disorder
Bunions	Flat feet	Osteoporosis
Cancers (only certain ones)	Glaucoma	Plantar fasciitis
Carpal tunnel syndrome	Gout	Polycystic ovarian syndrome
Cavities	Hammer toes	Pre-eclampsia
Chronic fatigue syndrome	Hemorrhoids	Rickets
Cirrhosis	High blood pressure (hypertension)	Scurvy
Constipation (chronic)	Iodine deficiency (goiter/cretinism)	Stomach ulcers
Coronary heart disease	Impacted wisdom teeth	
Crohn's disease	Insomnia (chronic)	

Source: After Lieberman 2013, Table 3, p. 173.

The different consequences of undirected, unpredictable environmental and cultural change

When we construct our environment, we change it in a certain direction and can predict many of the short-term changes. However, individual decisions and actions often accumulate to have population-level consequences that are difficult to anticipate. Global travel spreads both emerging diseases and bacterial genes for antibiotic resistance. Individual increases in energy consumption drive climate change, but so gradually and with such indirect causation that it is taking precious decades first to understand the process and then even more decades for nations to react to that understanding. Some consequences of such changes can eventually be managed, but the delay in response, measured in decades or centuries, results in much suffering and loss of life. Some threats, including some of the emerging diseases, may be very difficult to manage at all. One change in our environment with major consequences for health and disease has been the effect of our niche construction on our microbiota.

The roles of microbiota in human health

As we discussed in Chapter 5, adaptation to many aspects of the environment, including nutrients, toxins, and pathogens, depends not only on our own genes, but also on the microbial symbionts that colonize our intestines, skin, upper respiratory system, and genitourinary tract. These commensal organisms, which include bacteria, viruses, and fungi, play many roles; some roles are beneficial, and some are harmful. They induce the normal development of the immune and gastrointestinal systems in newborn infants; they are necessary for the normal functioning of the immune system; they synthesize some vitamins; and they mediate the digestion and absorption of nutrients, particularly fats and carbohydrates. The list of important organismal functions in which they are involved is already long and is growing rapidly. When those functions are disrupted by cultural changes in hygiene, diet, and medical practice, serious consequences and unpleasant surprises sometimes arise. It is not only our bodies that are mismatched to modern environments; our microbiota are mismatched as well.

■ SUMMARY

Mismatch describes both a poor fit of an organism to its environment and a situation in which selection acts to improve that fit. Many of the current mismatches that cause health problems occur because biology changes much more slowly than culture. Recent culture-driven changes include much less exposure to microorganisms, more artificial light, better temperature-controlled housing, profound changes in diet, much less physical activity, and more exposure to addictive drugs than we experienced in our evolution. Mismatch to these changes is probably responsible for at least 50 diseases and other health issues, many of which are related to changes in our microbiota.

Mismatches in time: The major cultural and epidemiological transitions

The mismatch that we currently experience has in part accumulated fairly steadily, but much of it is the product of two dramatic episodes that each greatly increased the rate of cultural evolution: the agricultural revolution, which started about 10,000 years ago, and the industrial revolution, which started about 200 years ago. Each produced major changes in diet, lifestyle, and our microbiota (Figure 8.1).

Prior to the invention of agriculture, when we were living as hunter-gatherers in small groups, we were regularly exposed to worms, *Helicobacter*, *Salmonella*, and lactobacilli, organisms with which our immune systems have shared a long coevolutionary history. We ate fruits, vegetables, nuts, and tubers that were high in fiber and often tough to chew as well as lean meat and fish. With agriculture and settlement in towns and cities, our populations increased in size and density; our diets changed toward an increased proportion of grains and other starches; we were exposed to farm animals, feces, mud, and untreated water; and we acquired new infectious

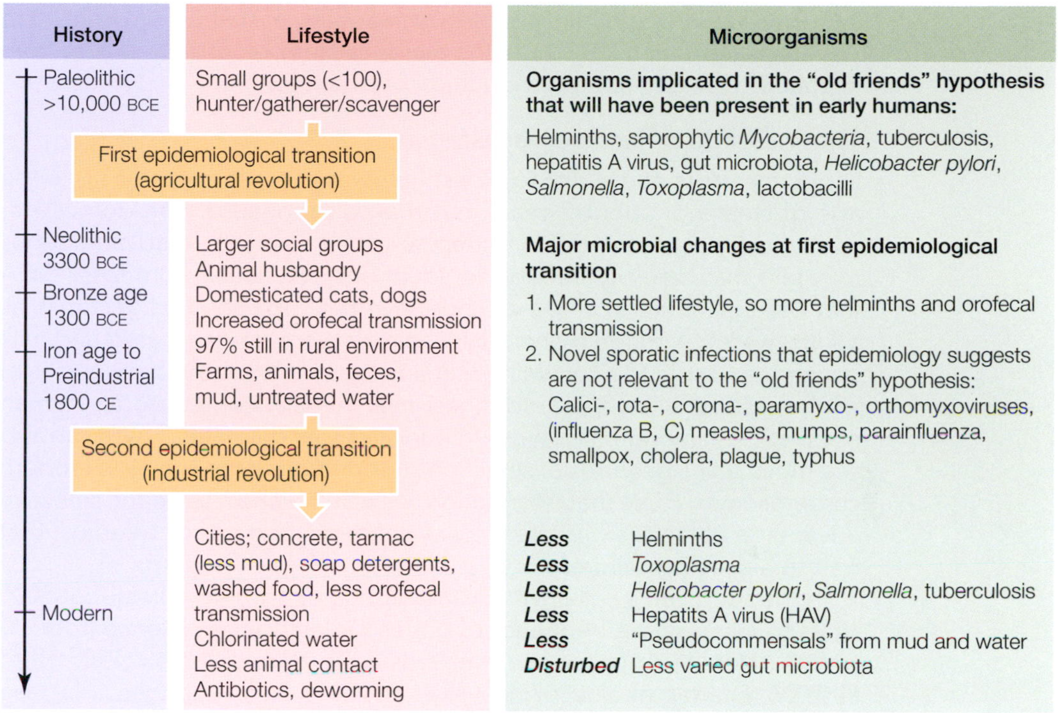

History	Lifestyle	Microorganisms
Paleolithic >10,000 BCE	Small groups (<100), hunter/gatherer/scavenger	**Organisms implicated in the "old friends" hypothesis that will have been present in early humans:** Helminths, saprophytic *Mycobacteria*, tuberculosis, hepatitis A virus, gut microbiota, *Helicobacter pylori*, *Salmonella*, *Toxoplasma*, lactobacilli

First epidemiological transition (agricultural revolution)

History	Lifestyle	Microorganisms
Neolithic 3300 BCE; Bronze age 1300 BCE; Iron age to Preindustrial 1800 CE	Larger social groups; Animal husbandry; Domesticated cats, dogs; Increased orofecal transmission; 97% still in rural environment; Farms, animals, feces, mud, untreated water	**Major microbial changes at first epidemiological transition** 1. More settled lifestyle, so more helminths and orofecal transmission 2. Novel sporatic infections that epidemiology suggests are not relevant to the "old friends" hypothesis: Calici-, rota-, corona-, paramyxo-, orthomyxoviruses, (influenza B, C) measles, mumps, parainfluenza, smallpox, cholera, plague, typhus

Second epidemiological transition (industrial revolution)

History	Lifestyle	Microorganisms	
Modern	Cities; concrete, tarmac (less mud), soap detergents, washed food, less orofecal transmission; Chlorinated water; Less animal contact; Antibiotics, deworming	*Less*	Helminths
		Less	*Toxoplasma*
		Less	*Helicobacter pylori*, *Salmonella*, tuberculosis
		Less	Hepatits A virus (HAV)
		Less	"Pseudocommensals" from mud and water
		Disturbed	Less varied gut microbiota

Figure 8.1 Two major cultural transitions, the agricultural revolution and the industrial revolution, have profoundly changed our relationships to disease. (After Rook 2012.)

diseases, many of which emerged from domesticated animals. Included were diseases caused by caliciviruses, rotaviruses, coronaviruses, paramyxoviruses, and orthomyxoviruses, such as influenza, measles, mumps, and smallpox, as well as bacterial diseases like cholera, plague, and typhus. Our evolutionary exposure to these diseases has not been as long as our exposure to parasitic worms and *Helicobacter pylori*.

The mismatch diseases that followed from the agricultural revolution include the nutritional disorders that result from a loss of diversity in our diets, such as scurvy (insufficient vitamin C), pellagra (insufficient vitamin B_3), beriberi (insufficient vitamin B_1), goiter (insufficient iodine), and anemia (insufficient iron), all now treated with dietary supplements. They also include tooth cavities, which are caused by increased starch and sugar, and some of the increase in type 2 diabetes, caused in part by the repeated rapid surges in blood sugar that occur after meals of refined starches and processed sugars.

The mismatch disorders that followed from the industrial revolution include a further increase in type 2 diabetes, obesity, cardiovascular disease (both heart attacks and stroke), osteoporosis, and colon cancer. The health consequences of the industrial revolution were by no means all bad, however. Among white Americans, infant mortality fell from 22% in 1850 to

0.6% in 2000, life expectancy at birth increased over that period from 40 to 77 years (Lieberman 2013), and the major causes of mortality shifted from infectious diseases to noncommunicable diseases.

The message of lactase persistence

In our ancestors, the ability to digest milk was present in infants up to the age of weaning, about 2 years, but then disappeared. The domestication of dairy animals—goats, sheep, and cattle—made it advantageous to retain that capacity through to adulthood. We know from fossil DNA that preagricultural humans did not have the capacity to digest milk as adults; we know roughly when dairy animals were domesticated and milk became available to nonnursing humans; and we know the frequencies of the genes that allow adults to digest milk in many contemporary human populations, among which those frequencies vary dramatically. We can also see molecular signatures of selection surrounding those genes in modern chromosomes. From that information, we can infer how large the selective advantage of lactase persistence was and how long it took to adjust that trait to a dairying culture.

In theory, given a selective advantage of 5%, it would take about 8000 years to increase the frequency of genes for lactase persistence from 1% to 90%. Neolithic history, however, was a bit more complex (Tishkoff et al. 2006; Enattah et al. 2007; Heyer et al. 2011). Local changes in lactase persistence were driven both by local selection and by immigration from areas where lactase persistence had already started to evolve. Depending on the region considered and the level of immigration assumed, the selective advantage of drinking milk in adulthood must have been between 0.5% and 3% to account for the patterns seen today and in the fossil record. Only a few populations are now at least 90% lactose tolerant. Lactase persistence originated several times, independently, in dairying cultures in Europe, Asia, and Africa. Milk was being consumed in Anatolia around 8500 BP (before present), in Britain around 6100 BP, and in Scotland around 3000 BP. In Africa, lactase persistence originated in nomadic cow herders around 7000 BP. The current prevalence of primary lactose tolerance ranges from 95%–97% in Scandinavia to 0%–10% in East Asia.

The message of the lactase example is important: evolution takes hundreds of generations to adjust populations to new conditions. Strong selection and 8500 years were not enough to fix the trait. Mismatch is not just plausible; in cases like this one, it is real. It took a great deal of research to establish that point.

■ SUMMARY

Both the agricultural and the industrial revolutions produced major shifts in lifestyle, diet, and exposure to microorganisms to which our biology has not yet fully adjusted. During the agricultural revolution we acquired influenza, measles, smallpox, cholera, plague, and typhus, among other infectious diseases, as well as nutritional disorders. During the industrial revolution we experienced increases in obesity, type 2 diabetes, osteoporosis, cardiovascular disease, and colon cancer, as well as lower infant mortality and longer life. One major shift in diet caused by the agricultural revolution was dairy milk consumed by adults. The fact that our genomes have not yet fully adjusted to milk in our diet makes clear the point that biology changes more slowly than culture.

Mismatches in space: Emigration and immigration

Migration creates mismatches in at least three ways. Migrants introduce new diseases to populations that had not previously experienced them, immigrants leave some pathogens and symbionts behind, and immigrants switch to novel diets and lifestyles.

When Columbus's sailors returned to Europe from the New World, they brought syphilis with them. In Europeans, it was initially a very virulent disease, killing most of those infected within a few weeks. When syphilis spread out of Naples, it routed the French army, which was then invading Italy. In return, Europeans introduced measles, smallpox, bubonic plague, influenza, typhus, diphtheria, and scarlet fever to the New World and Polynesia, where these diseases killed from 20% to more than 90% of the natives. The impact of smallpox was particularly impressive (Table 8.2).

Immigrants also often leave pathogens, vectors, and symbionts behind them. The slave trade brought Africans with sickle-cell genes to North America. When malaria was eradicated in North America, the compensating advantage that had existed in Africa disappeared, and these Africans were left with the costs of a genetic disease that produces serious anemia. Other immigrants suffer the consequences of changes in their microbiota. First-generation immigrants born in developed countries often suffer from allergies and asthma at higher rates than do children born in the developing world but living in the same communities in the developed world. For example, international adoptees and children born in Sweden to foreign-born parents used asthma medication three to four times as often as did foreign-born children whose immune systems had been conditioned early

TABLE 8.2 Smallpox epidemics

Location	Year(s)	Number of deaths
Hispaniola	1507 and 1518	Up to 300,000
Mexico	1520–1521	2 million–15 million
Peru	1525–1527	200,000
Brazil	1555–1878	3 million
Venezuela	1580	Up to 30,000
Connecticut	1634	Up to 90%
Massachusetts	1617–1649	Up to 90%
Iroquois Confederacy	1662	Up to 20,000

Source: After Kohn 2008.

in life by exposure to a broader range of infections and symbionts (Bråbäck, Vogt, and Hjern 2011).

Immigrants also switch to new diets and lifestyles, and that can have mixed results. Japanese immigrants to Hawaii switched from a diet high in fish, rice, and tea to one that contained more red meat, animal fat, and milk. One result was that men of Japanese ethnicity living in Hawaii were twice as likely to die of heart disease as were Japanese men living in Japan (Robertson et al. 1977); another was that the children of the immigrants were on average about 10 cm taller than their parents (Froehlich 1970).

■ SUMMARY

Mismatches occur across space as well as through time. The movement of humans and their diseases across the planet exposes some people to new diseases while allowing others to leave microorganisms behind, both those that cause disease and those that are beneficial. Immigrants also experience changes in diet and microbiota that can dramatically change the prevalences of noncommunicable diseases.

Obesity, a condition with many causes

People with a body mass index (BMI) between 25 and 30 are considered overweight; those with a BMI higher than 30 are considered obese. In 1900, the *average* BMI of American men 40–59 years old was 23; by 2000, it was 27.5. In 2008, more than 15% of the adolescent population of the United States was obese and more than 33% was overweight, problems that can later lead to hypertension, type 2 diabetes, cardiovascular disease, and social stigma (Metzger and McDade 2010). Mismatch contributes to obesity in many ways: through energy balance, food quality, changes in amounts and patterns of sleep, changes in microbiota resulting from less breast-feeding and more births by cesarean section, and interactions between undernourishment of fetuses and infants and overnourishment of adults.

Energy balance: Too much food, too little exercise

The most important of these multiple causes is a major recent shift in energy balance: we are now more sedentary than we once were, and we have easy access to more food than is good for us. Both trends started with the agricultural revolution, have been intensified by the industrial revolution, and can be countered by eating less and exercising more. They can be quantified in estimates of changes in physical activity level (PAL), the ratio of total daily energy output to basal metabolism. In hunter-gatherers, PALs average 1.85, which is about the same as those of farmers (1.78–1.86) but considerably higher than office workers (1.56–1.61); if a hunter-gatherer or farmer changes careers and becomes an office worker while maintaining the diet of 3000 calories per day that a farmer needs, he will have an energy surplus of about 450 calories per day and will, unless he compensates with exercise, become obese (Lieberman 2013).

TABLE 8.3   Comparison of hunter-gatherer and American diets with U.S. recommended daily allowances

Item	Hunter-gatherer	Average American	U.S. recommended daily allowance
Carbohydrate (% daily energy)	35%–40%	52%	45%–65%
Simple sugars (% daily energy)	2%	15%–30%	<10%
Fat (% daily energy)	20%–35%	33%	20%–35%
Saturated fat (% daily energy)	8%–12%	12%–16%	<10%
Unsaturated fat (% daily energy)	13%–23%	16%–22%	10%–15%
Protein (% daily energy)	15%–30%	10%–20%	10%–35%
Fiber (g/day)	100 g	10–20 g	25–38 g
Cholesterol (mg/day)	>500 mg	225–307 mg	<300 mg
Vitamin C (mg/day)	500 mg	30–100 mg	75–95 mg
Vitamin D (IU/day)	4000 IU	200 IU	1000 IU
Calcium (mg/day)	1000–1500 mg	500–1000 mg	1000 mg
Sodium (mg/day)	<1000 mg	3375 mg	1500 mg
Potassium (mg/day)	7000 mg	1328 mg	580 mg

Source: After Lieberman 2013, Table 5, p. 224

Food quality

The shift in energy balance that accompanies a more sedentary lifestyle is certainly one important reason for obesity, but it is not sufficient to explain the dramatic recent increase in the problem. Another major contributor is the recent shift to diets based on industrial foods that emphasize simple starches and refined sugars (Table 8.3); they are nutrient-poor and calorie-rich.

The spike in blood sugar that results from ingesting simple starches and refined sugars—high glycemic foods—stimulates the pancreas to produce insulin, which causes the sugar in the blood to be taken up by skeletal muscle and adipocytes. The unnatural spike is so rapid, however, that the pancreas often produces too much insulin, causing blood sugar levels to crash and leading to a rapid return of hunger and overeating that does not occur when complex starches and sugars associated with fiber are ingested (Lieberman 2013).

Sleep deprivation and ghrelin

We are also stimulated to overeat by changes in our sleep patterns. Representative hunter-gatherers arise at dawn, take a 2-hour nap in the afternoon, and are asleep by 9 p.m. A typical modern American sleeps only 6.1 hours per night, and only a third take naps (Lieberman 2013). Lack of sleep reduces the normal pulse of growth hormone that supports cell repair and immune function and increases the secretion of cortisol, which raises the level of blood sugar. When high levels of cortisol persist, they can depress immune responses, dampen growth, and increase the risk of type 2 diabetes. Sleep deprivation also causes levels of leptin, which inhibits appetite, to fall and levels of ghrelin, which stimulates appetite, to rise, leading sleep-deprived people to be hungry and to crave carbohydrates in particular (Lieberman 2013). Evolution shaped our bodies to interpret sleep deprivation as a type of stress that needed to be managed with increased food intake.

Breast-feeding and cesarean sections

Also contributing to the risk of obesity is the switch from breast-feeding to infant formula, perhaps mediated by a shift in that portion of our gut microbiota that helps digest carbohydrates and lipids. When controlled for other factors, the average weights of offspring of mothers who breast-fed were 13 pounds lighter at 14 years of age than were the weights of the offspring of the siblings of those mothers who did not breast-feed (Metzger and McDade 2010). Delivery by cesarean section also contributes to the risk of obesity. One study of 1255 children in Boston suggests that birth by cesarean section roughly doubles the risk of obesity at age 3 when compared with vaginal birth (Huh et al. 2012). The intestinal microbiota of infants born by vaginal delivery is similar to the vaginal and fecal microbiota of their mothers, but the microbiota of infants delivered by caesarian section is more similar to that of their mothers' skin. Whether this difference has an effect on long-term metabolic programming remains to be established.

Abnormal weight early in life

Finally, as discussed in Chapter 2, the nutritional status of the fetus in utero and of infants and young children also produces a delayed effect on the risk of adult diseases. Infants and children who are undernourished develop into adults with greater risks of obesity, type 2 diabetes, high blood pressure, and cardiovascular disease—a collection of illnesses known as the metabolic syndrome—than do those who are well nourished. One explanation for this effect is mismatch: the environment of the young organism is mismatched to the environment of the adult, and it triggers a lifelong state of insulin resistance that becomes an inappropriate reaction to an environment that is subsequently energy-rich. Such a reaction has high costs later in life. One reason it might have evolved is that it is critically important to protect brain development in infancy and childhood with adequate supplies of blood sugar, and one mechanism for doing that is to make the rest of the body more insulin resistant, slowing the uptake of sugar in nonbrain tissues.

On the other hand, infants and children who are overweight also have an increased risk of obesity as adults. The mechanisms that mediate this risk are complex. They may involve greater expansion of the visceral fat mass that can lead to stable changes in appetite and body-weight set points that perpetuate overeating and promote obesity. Adipose tissue is the source of leptin, a hormone that controls appetite and energy expenditure. The level of leptin circulating in the blood is proportional to fat mass; thus, a high level of leptin indicates that sufficient energy stores are available. However, the persistent presence of very high levels of leptin results in leptin resistance. When that happens, the hunger and satiety centers in the hypothalamus become less responsive to leptin, leading to increased appetite and decreased energy expenditure. This vicious cycle contributes to lifelong obesity and its resulting metabolic diseases.

■ SUMMARY

For obesity, as for other mismatch diseases, the causes are many and the treatments are correspondingly numerous. People become obese because they eat too much and exercise too little, because they eat processed foods that trigger insulin spikes that make them hungry sooner, because some of them have particular genes, because they are sleep deprived and produce more ghrelin, because their microbiota may have been disturbed by a lack of breast-feeding or by cesarean delivery, or because they were undernourished as infants and children.

Type 2 diabetes

The proximate cause of type 2 diabetes is insulin resistance, a condition that arises through an interaction between genes and environments. Let's first look at what usually happens. When blood glucose rises after a meal, it moves into cells through glucose transporters in the cell membranes. The glucose transporter GLUT2 is expressed in the liver and the endocrine pancreas and only transfers glucose when the plasma level of glucose is very high. The glucose transporter GLUT3 is expressed in the brain and transfers glucose even when the plasma level of glucose is low, thus ensuring that the brain has higher priority for glucose and oxygen consumption than most other tissues. The glucose transporter GLUT1 operates in most cell types, where it is expressed at a level corresponding to the functional demands of the corresponding tissue. For example, the level of GLUT1 is increased in T cells when they are activated, adjusting the fuel supply to their energy needs. The fourth glucose transporter, GLUT4, is expressed in skeletal muscle and fat, where it transports glucose in an insulin-dependent manner, in contrast to other tissues, in which glucose transport is insulin-independent. Normally, the pancreas releases insulin when blood sugar rises, and fat and muscle cells then take up glucose. This glucose is either used immediately for energy (in exercising muscle) or is converted into glycogen or fat in liver, muscle, and adipose tissue.

Because skeletal muscle and fat are the major consumers of glucose, insulin essentially controls glucose allocation among tissues. Allocation priorities change, for example, during pregnancy or infection, when the glucose supplies to the fetus and to the immune system become more important than the glucose supply to muscle and fat. That shift in allocation is orchestrated in part by reducing the glucose sensitivity of muscle and fat: they become insulin resistant.

Insulin resistance in muscle, fat, and liver cells can also occur under pathological conditions associated with obesity and chronic inflammation. When the ability of adipose tissue to store lipids is overwhelmed or otherwise impaired by obesity, the overflow of fatty acids can cause lipotoxicity because most cell types cannot accumulate lipids without deleterious consequences. When fatty acids accumulate in skeletal muscle, they trigger signaling pathways that promote insulin resistance. Inflammatory cytokines, whose release is not only triggered by infection but are also secreted by adipose tissue, can also cause insulin resistance both directly, by blocking insulin signaling in fat, muscle, and liver cells, and indirectly, by blocking lipid deposition in adipose tissue, thus leading to an increase in circulating fatty acids and their accumulation, as described above, in skeletal muscle and other tissues.

Exercise effectively prevents and can even reverse the early stages of insulin resistance in at least two ways. It both promotes metabolism of excess fatty acids in skeletal muscle, and it activates insulin-dependent glucose uptake by skeletal muscle. Conversely, insulin resistance can result from chronic lack of exercise and from diets rich in rapidly digestible glucose and fructose, both elements of modern industrial foods that were much less available in preindustrial diets. Thus, fuel allocation controlled by insulin evolved to be flexible to meet varying environmental challenges and the corresponding changes in physiological priorities. The system can be derailed by two elements of modern environments: industrial foods and lack of adequate exercise. In many cases, diet and exercise alone can reverse type 2 diabetes, particularly if it is caught at an early stage, and can be more effective than drug treatments (Lieberman 2013).

An individual's recent history and current condition are not the only contributors to the risk of type 2 diabetes. As we saw above for obesity, the risk of developing the metabolic syndrome later in life is influenced by experiences encountered by the fetus during pregnancy and by the infant and child early in life. Infants and children who are undernourished and below the normal weight for their height develop insulin resistance that persists throughout their lives, making them more likely to develop type 2 diabetes.

As with obesity, there is genetic variation for risk of type 2 diabetes. Genetic variants contributing to that risk have effects mediated by the environment, particularly exercise and diet.

■ SUMMARY

Type 2 diabetes is a mismatch disease caused by an increase in insulin resistance in fat and muscle cells. Insulin resistance has causes that include obesity-related inflammation, lack of exercise, a diet rich in glucose and fructose, and poor nourishment as an infant or child. Type 2 diabetes is a disease of homeostasis ultimately caused by the use of a control system with adjustable set points that led to a vulnerability that mismatches exploit.

Cardiovascular disease

Cardiovascular diseases, such as atherosclerosis, are among the leading causes of mortality in industrialized countries. Athersclerosis is accompanied by the thickening of arterial walls due to the accumulation of macrophages and lymphocytes in the intima. The stages of atherosclerosis are not yet fully understood, in part because the processes that produce the full-blown disease can take decades to unfold. The key ingredients are inflammation, macrophages, and lipid metabolism, particularly the metabolism of cholesterol.

Cholesterol and triglycerides are distributed throughout the body by the circulation of lipoprotein particles. Cholesterol is transported from the liver to other tissues by low-density lipoproteins (LDLs), and excess cholesterol is transported from the tissues back to the liver by high-density lipoproteins (HDLs). Because animals cannot catabolize cholesterol, it must be converted to bile acids to be excreted through the intestine.

Normally, the cholesterol level in cells and tissues is tightly controlled to prevent the toxic effects of cholesterol accumulation. When cells have enough cholesterol for their biosynthetic needs, they turn down the expression of the LDL receptor that governs LDL and cholesterol uptake. However, inflammation can produce oxidized LDLs (oxLDLs) that promote further inflammation and can be taken into macrophages by scavenger receptors. Unlike the LDL receptor, scavenger receptors are not down-regulated when the concentration of cholesterol in macrophages is high. Instead of leading to the normal negative feedback control of homeostasis, the increased uptake of oxLDLs produces more inflammation, increasing the recruitment of macrophages to the intima, increasing the permeability of blood vessels, and thus increasing access to LDLs circulating in the plasma, which produces more oxLDLs. This vicious cycle can operate at low levels for many years, producing progressively enlarged atherosclerotic plaques. When those plaques become large enough, they lead to alterations in the vasculature and to clotting. These conditions progress to dangerous late-stage atherosclerosis, when plaque rupture can cause myocardial infarction or stroke.

The factors in the environment that raise the risk of cardiovascular disease by increasing the bloods levels of LDLs and lowering those of HDLs include physical inactivity, poor diet, obesity, drinking, smoking, and emotional stress. Moderate exercise increases levels of HDLs and lowers blood pressure by stimulating growth of new blood vessels and strengthening heart muscle. A diet rich in unsaturated fats increases HDLs; one rich in saturated fats increases LDLs. Too much alcohol damages the liver, degrading its ability to regulate the levels of HDLs and LDLs in the blood. Smoking both damages the liver and inflames the walls of arteries. Emotional stress increases blood pressure and adrenaline secretion with direct effects on cardiovascular physiology and can lead to behavioral changes, including alcohol consumption, with indirect effects on the risk of heart disease. Although cardiovascular disease, like obesity and type 2 diabetes, does have a genetic component, the genetic risks are expressed through environmental interactions. Cardiovascular diseases are largely mismatch diseases caused by sedentary lifestyles, diets rich in postindustrial processed fats and sugars, and two novelties to which we do not have long evolutionary exposure: alcohol and nicotine (Lieberman 2013), which are rare in hunter-gatherers.

■ SUMMARY

Cardiovascular disease is a mismatch disease with multiple causation mediated by inflammation, macrophages, and cholesterol metabolism. Physical inactivity, poor diet, obesity, drinking, smoking, and emotional stress all contribute to changes in cardiovascular physiology that produce the growth of plaques whose rupture causes myocardial infarction or stroke.

Female reproductive cancers

As we saw in Chapter 6, every cancer is an independent instance of clonal evolution fueled by genetic variation generated in large part by somatic mutations. Mutations occur when cells divide, and cells divide in the female reproductive tract, particularly in breasts and ovaries, during every menstrual cycle. Female hunter-gatherers spent much more of their lives pregnant and nursing than do women in postindustrial societies. Nursing represses ovulation and menstruation, a condition called lactational amenorrhea. As a result, whereas hunter-gatherer females, like our presumed ancestors, experience about 150 menstrual cycles per lifetime, women in postindustrial societies experience 350 to 400 menstrual cycles, which implies nearly three times as many opportunities for somatic mutations to occur in dividing cells in breasts and ovaries (Strassmann 1999). Obesity also increases the risk of reproductive cancers because fat cells release estrogen into the bloodstream, causing reproductive cells to divide. In postmenopausal American women, obesity increases the risk of breast cancer 2.5 times (Lieberman 2013).

Although the number of menstrual cycles per lifetime is influenced by the use of oral contraceptives, with more cycles in women who use them, the effect of oral contraception on cancer risk is complex. The use of oral contraceptives increases the risk of breast cancer but decreases the risk of ovarian cancer, resulting in little net change in mortality risk (American Society for Reproductive Medicine 2008).

■ SUMMARY

The risk of female reproductive cancers is affected both by oral contraception and by obesity. In our evolutionary past the female body normally experienced about 150 menstrual cycles per lifetime; women using oral contraceptives can now experience 350 to 400 cycles per lifetime. In every menstrual cycle cells divide, and in every cell division there is a risk of a mutation that could lead to cancer. Obesity also contributes to cancer risk because it increases estrogen secretion that causes cell division in estrogen-sensitive tissues.

Hygiene and old friends, asthma, and autoimmune diseases

With industrialization came modern cities with soap, washed food, less mud than in the countryside, and chlorinated water, which resulted in less oro-fecal transmission and less contact with animals. Antibiotics reduced our exposure to all types of bacteria, not just the pathogenic ones, and antihelminthics reduced our exposure to parasitic worms. We now also encounter less *Helicobacter*, *Mycobacterium tuberculosis*, *Salmonella*, and *Toxoplasma* as well as less hepatitis A virus than previously. Our microbiota is less varied than it once was.

Correlations between parasite exposure, atopies, and autoimmune diseases

As infectious diseases have declined in the postindustrial world, autoimmune diseases and atopies have increased (see Chapter 3). That is a correlation in time. There are equally striking and suggestive correlations in space. Doctors in the tropics do not see much autoimmune disease or allergy. Autoimmune diseases like type 1 diabetes are found in countries in which worms, leprosy, and tuberculosis are rarely encountered (Figure 8.2). Such correlations in time and space between exposure to parasites and pathogens and the incidence of asthma, allergies, and autoimmune

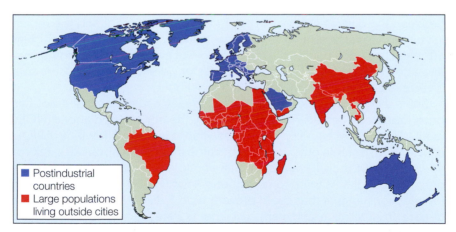

■ Postindustrial
countries
■ Large populations
living outside cities

Figure 8.2 Postindustrial countries (in blue) have high rates of type 1 diabetes (>8 per 100,000 per year) and few parasites; countries that have large populations who live outside cites or who otherwise lack access to clean water and antibiotics (in red) have very low rates of type 1 diabetes and six or more of these parasites and diseases: filariasis, onchocerciasis, schistosomiasis, soil-transmitted helminths like hookworm, trachoma, leprosy, and many other diseases. (After Zaccone et al. 2006.)

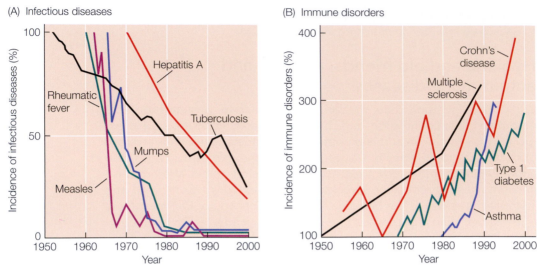

Figure 8.3   There is an inverse relationship between infectious diseases (A) and immune pathologies (B). (After Bach 2002.)

diseases suggested the hygiene, or old friends, hypothesis (Figure 8.3). The old friends mentioned in the hypothesis are the symbiotic species in our microbiota with which we have had a long coevolutionary history to which our immune systems adjusted (Rook 2012). It is their loss from our microbiota that is thought to trigger some of the inappropriate immune responses that produce asthma, allergies, and autoimmune diseases.

Many potential mechanisms might be mediating the relationship between the diverse changes in our modern environment and the recent increases in cases of asthma, allergies, and autoimmune diseases. Particular attention has been paid to the possible role of parasitic worms; somewhat less attention has been given to the roles of general early exposure to a diverse microbiota, of diet, and of artificial light.

Some evidence suggests that worms do play a role. Worm therapy has been repeatedly tested in a mouse model, the nonobese diabetic (NOD) mouse (Zaccone et al. 2006). Initial observations indicated that 50% of NOD mice reared in sterile environments became diabetic at about 20 weeks of age, whereas it took 40 weeks for 50% of mice reared under conventional breeding conditions to become diabetic. That finding prompted a search for infectious agents that could prevent type 1 diabetes, an autoimmune disease, in NOD mice, and many such agents were found. The program was then extended to animal models of other autoimmune diseases, including inflammatory bowel disease, and again, several worms or worm products were found that did prevent autoimmunity (Table 8.4).

Based on those results in model systems, a search was carried out for a therapeutic helminth (Zaccone et al. 2006). Such a worm should have little or no pathogenic potential; it should not multiply in the host; it should not spread to other hosts; its colonization of humans should be self-limited and

TABLE 8.4 Worms or worm products that prevent autoimmunity in animal models

Agent or product	Autoimmune disease
Schistosoma mansoni	Experimental autoimmune encephalomyelitis, Grave's thyroiditis
S. mansoni eggs	Experimental autoimmune encephalomyelitis, experimental colitis
Trichinella spiralis	Experimental colitis
Trichuris suis	Inflammatory bowel disease
Heligmosomoides polygyrus	Experimental colitis
ES-62 (*Acanthocheilonema viteae* product)	Collagen-induced arthritis

Source: After Zaccone et al. 2006, Table 4, p. 521.

asymptomatic; its action should not be affected by commonly used drugs; it should be easily eradicated with antihelminthic drugs; it should be capable of being isolated in pure form, without other pathogens, in large numbers; it should be capable of being stabilized for transport and storage; and it should be easy to administer. Those criteria eliminated virulent pathogens. The leading remaining candidate was the pig whipworm, *Trichuris suis*. It is a close relative of the human whipworm, *T. trichiura*, which can establish an enduring infection in the human gut that is pathogenic enough to require treatment. In humans, the pig whipworm does establish an infection, but it is not very pathogenic, and symptoms, if any, are mild.

Subsequent preliminary clinical trials used cocktails of *T. suis* eggs, either administered once or in repeated doses (Elliott, Summers, and Weinstock 2007). Although some individuals experienced improvement, the overall results were mixed. For example, in a double-blind, placebo-controlled trial of 54 patients with active ulcerative colitis, 43% of those treated with *T. suis* eggs improved, which was more than the 17% who improved on the placebo. These results encouraged larger, longer clinical trials, two of which (see below) were called off in 2013 when those on treatments did no better than those on placebos (see clinicaltrials.gov for recent updates).

Case control studies have also been done. One is particularly interesting because the autoimmune disease in question is multiple sclerosis (MS), a potentially devastating disease for which treatment has been difficult. In this study (Correale and Farez 2011), all the patients admitted had started to develop symptoms of MS. They were divided into two groups: those naturally infected with helminths and those who were uninfected. The patients were then followed for 7.5 years, with multiple measures of the progress of the disease using brain scans and other techniques made every 6 months. Uninfected patients developed the brain lesions characteristic of MS, called T2 lesions, much more rapidly than did the infected patients. After 5 years, some of the patients who were infected with worms were

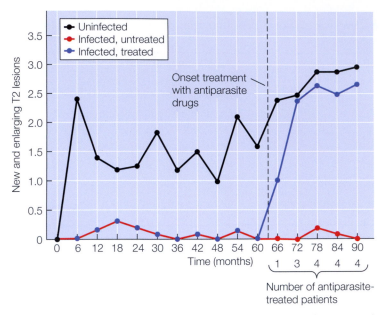

Figure 8.4 Parasitemia slows the progress of multiple sclerosis, and treating the parasites accelerates it. By 66 months, one patient, by 72 months, three patients, and after 78 months, four patients were being treated for worms. The points in the graph are mean values. (After Correale and Farez 2011.)

suffering from their worm infections and were treated with drugs to clear the infections. That group of treated patients then rapidly developed T2 lesions (Figure 8.4).

How might worms be interacting with the immune system?

The mechanisms by which worm infections might prevent or suppress the autoimmune response are not known. Several scenarios are possible. First, the immune response produced by worm infections overlaps with the tissue repair response (Gause, Wynn, and Allen 2013). Second, parasitic worms have evolved many ways to manipulate the immune system, including potent immune-suppressive mechanisms. Therefore, some aspects of the immune response may have been globally suppressed in the past by the immune evasion signals produced by worms; when the worms are removed, the immune system, unaccustomed to their absence, may become hypersensitive to stimulation. Third, some aspects of immune responsiveness may be programmed by infections early in childhood that affect both the thresholds of responsiveness and the types of responses induced (e.g., antibacterial vs. antihelminthic). These and other possibilities are all supported by some evidence, but none of them have been demonstrated directly, nor have any been ruled out.

Worm therapy is a risky trade-off

In the tropics, worm infections increase the susceptibility of human hosts to tuberculosis, AIDS, malaria, and other infectious diseases. They also reduce the protection offered by vaccines by interfering with the immune response to vaccination. As we saw in the study of worm-infected patients with MS (Correale and Farez 2011), the symptoms caused by the worms can become so serious that removing the worms with drug therapy may be advisable even though it results in a distinct worsening of MS symptoms. Some individuals have reported improvements in the symptoms of inflammatory bowel disease after acquiring hookworm infections, but hookworm infections produce serious pathology and are a cost not to be taken lightly.

Clinical trials of worm therapy have not yet been successful

Although a phase 1 clinical trial suggested that worm therapy for inflammatory bowel disease might work, Coronado Biosciences, which was conducting a phase 2 clinical trial of pig whipworm therapy for Crohn's disease, announced in October 2013 that it was calling off the trial because patients were not doing better with the therapy than with a placebo. A parallel trial in Germany run by the company Dr. Falk Pharma was called off a month later for the same reason. It may be that worm therapy is only effective if worms that can establish enduring infections and cause serious pathogenic effects are used, calling into question the cost-benefit balance of the treatment.

■ SUMMARY

Our bodies have coevolved with a diverse community of symbiotic and pathogenic bacteria and parasites. Many of our interactions with them are complex, having both positive and negative aspects. When sanitation, hygiene, or drug therapy removes those commensals, our immune systems respond inappropriately, producing allergy, asthma, eczema, and autoimmune diseases. The mechanisms that mediate those responses are now partially understood, but therapies using biotic agents have not yet been successful.

Conclusion

There is no question that our bodies are mismatched to many aspects of our modern environment, most importantly to a sedentary lifestyle, a postindustrial diet, a lack of exposure to parasitic worms, and an altered exposure to microorganisms. Some of the consequences of these mismatches—obesity, high blood pressure, cardiovascular disease, and type 2 diabetes—can be improved with lifestyle interventions such as better diet and regular moderate exercise. For others, particularly the autoimmune diseases, effective therapies are desperately needed but not yet available.

Mental Disorders

Mental disorders are usually expressions of cognition and behavior that do not affect functionality in most individuals but are disadvantageous in a few. Cognition and behavior are traits with very complex causation. They are produced by interactions among genes and genetic control networks, developmental processes involving cascades of intercellular interactions, physiological processes involving hormonal regulation and homeostatic feedback, and interactions with the environment, including the family and other social environments. Experiences early in life and learned behaviors often play significant roles.

The special difficulties of explaining mental disorders

Processes with such complex causation do not yield easily to experimental analysis even where experiments are possible. In addition, many of the conditions described as mental disorders do not have credible equivalents in model organisms, putting experimental approaches effectively out of reach. We therefore do not yet have a solid understanding of the causes of many of the mental disorders or the degree to which evolution has directly or indirectly shaped our vulnerability to them. Because most of them, most of the time, have several to many causes, hypotheses are not mutually exclusive. We are faced with trying to estimate the relative contributions of a series of factors rather than the identification of an exclusive cause. That does not mean that particular mechanisms cannot play a leading role; several genetic and several environmental factors can all affect the same cellular mechanisms. It does mean, however, that claims

Figure 9.1 A null hypothesis for the frequency distribution of mental disorders in a population.

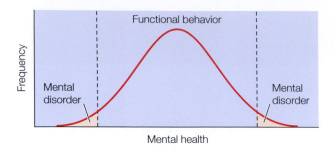

A null hypothesis

In discussing trade-offs in Chapter 2, we introduced the idea that "any complex process or structure influenced by several to many genes and several to many environmental factors, each of which may vary, will be expressed as a broad range of phenotypes, roughly as a bell-shaped curve." Here we take up that idea again in the context of mental disorders and advance it as a null hypothesis that explains the existence of abnormal cognition and behavior occurring in populations at low frequencies, often about 1%–3% (Figure 9.1). The existence of these disorders does not necessarily need any special evolutionary explanation; if they are simply the inevitable consequence of building complex traits out of parts some of which can vary for any reason, they will occur whether shaped by evolution or not. It is therefore not the existence of mental disorders per se that might benefit from evolutionary insights but, rather, particular features of particular disorders, which could carry a trace of past selection, trade-offs, or other evolutionary effects.

How evolution might enter the picture

Evolution is opportunistic in the sense that it usually exploits preexisting structures and mechanisms in shaping responses to new challenges. If a mutation produces a novel aspect of cognition or a change in behavior that increases reproductive success, it probably does so by modifying a process that had previously evolved for another purpose. When we combine that idea with the one depicted in Figure 9.1, we arrive at one strategy for exploring the evolutionary causation of mental disorders: we expect them to be associated with underlying mental structures that evolved for other reasons, structures whose evolutionary functions are associated with the variation normally expressed close to the center of the trait distributions, not with the extremes. Thus, we expect mental disorders to be correlated with other axes of variation that represent preexisting processes on top of which modifications of cognition and behavior evolved. Those correlations with other traits could well appear to be arbitrary in that they reflect the opportunistic way that selection explores phenotype space: it always works with what is available when a novel mutation with a fitness advantage arises, and that novel mutation arises at random, drawn from a much larger set of many possible mutations with many possible benefits and many possible side effects.

As is the case with other diseases, the causes of mental disorders can belong to several of the categories discussed in Chapter 3. Thus, rare monogenic disorders affecting brain development and function as well as trisomies result in mental retardation disorders, including Rett's syndrome, fragile-X syndrome, and Down syndrome. Mental retardation can also result from rare and severe environmental insults, such as poisoning, asphyxia, and trauma. Both of these groups of mental disorders are very indirect consequences of human evolution and development. As with other diseases caused by rare genetic or environmental catastrophic events, they do not have any obvious "normal" counterpart. On the other hand, some mental disorders can be viewed either as direct consequences, or as by-products, of design features, system constraints and vulnerabilities, mismatch, costs of defense, and genomic conflicts.

Mental disorders as by-products of defense systems

Many mental disorders can be viewed as exaggerated expressions of the corresponding normal behavioral, emotional, and cognitive traits. Cases in point include disorders of behavioral defenses, such as anxiety, phobias, and some forms of obsessive-compulsive disorders (OCDs). Humans have evolved several mechanisms to avoid infections. Olfactory detection of volatile microbial metabolites induces aversive behaviors that help minimize pathogen exposure, for example, and visual cues associated with infections similarly trigger feelings of disgust and promote contact avoidance. These normal behaviors provide obvious benefits. However, excessive expression of these defenses can manifest as abnormal germophobia characterized by avoidance of physical contact with other people or objects as well as excessive washing of hands and showering. Notably, most people express some degree of germophobia, and it is only the extreme forms of aversive behaviors that interfere with normal life that are pathological. Hoarding behavior is another example of OCD that has a normal counterpart, foraging behavior. Although foraging behavior may not be viewed as normal in the modern environment of industrialized countries, it presumably was normal in the evolutionary past. Thus, hoarding behavior may be an example of excessive expression of atavistic behavioral programs.

Other examples of mental disorders that may be by-products of behavioral defenses include anxiety, phobias (e.g., fear of open spaces), paranoia, and panic disorders. These disorders are excessive expressions of behavioral traits that normally promote avoidance of dangerous environments.

Mental disorders as diseases of homeostasis and mismatch

In our discussion of disease categories in Chapter 3, we defined category 3 as diseases of homeostasis often exposed by environmental mismatches. They are expressions of underlying systems that are designed to be adjustable, with set points defining the states that homeostasis seeks to maintain, and are therefore vulnerable to disregulation because disruption of the set points can produce extremes of expression. Some homeostatic mechanisms are behavioral, using internal systems of reward and punishment, pleasure and pain, to reinforce adaptive behavior and punish maladaptive behavior (Nesse 1994), and some mental disorders appear to be diseases of homeostasis exposed by environmental mismatches, including exposure to novel substances.

■ SUMMARY

Mental disorders are hard to explain because they have complex causation and credible model organisms are not available for experimentation. There is, however, a simple null hypothesis: some of them represent the rare tails of distributions of cognition and behavior whose centers are normal and selected. Mental disorders can result from rare, monogenic mutations, for example, trisomy 21. They can be byproducts of defense systems, such as obsessive-compulsive and panic disorders. And they can be diseases of homeostasis and mismatch.

Drug addiction

Drug addiction—compulsive drug use despite serious negative consequences—is clearly maladaptive. There are two main evolutionary hypotheses that try to explain why we are vulnerable to addictive behavior, and they are not mutually exclusive. The first, more thoroughly explored, hypothesis posits that drugs are evolutionary novelties that hijack innate systems that either evolved to regulate behavior by rewarding with pleasure behaviors that contribute to reproductive success, such as eating, sex, and nursing, or that adaptively suppress pain (Nesse 1994; Nesse and Berridge 1997). Under this hypothesis, drug addiction is a disease of homeostasis exposed by mismatch. The second hypothesis, which is more recent and less well explored than the first, posits that drug use originated through adaptive self-medication with plants with curative properties (Hagen et al. 2009).

There are two types of reward systems and two corresponding types of pleasure sensation. One has to do with the excitement and anticipation of satisfying some need, such as food consumption or sexual intercourse. This reward system operates even in the absence of actual satisfaction of the need that drives it. We feel happy in anticipation of consuming food when we are hungry and even when we have actually satisfied our hunger. A very different kind of pleasure sensation—satisfaction—is elicited by the actual consumption of food. The two types of pleasure sensations are mediated by distinct neuronal pathways: the dopaminergic and the opioid pathways, respectively. These pathways are targeted by different classes of addictive drugs. Thus, the psychostimulants cocaine and methamphetamine activate the dopaminergic pathway, whereas opium and heroin activate the opioid pathway. The flip sides of the two types of pleasure correspond to two types of negative emotional states: frustration, depression, and despair on the one hand and the feelings of hunger, thirst, and pain on the other. These

(A)

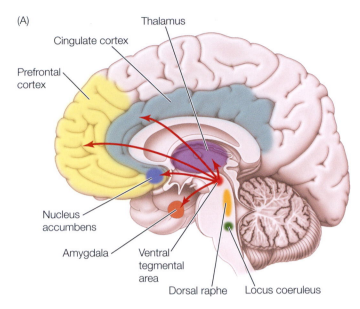

(B)

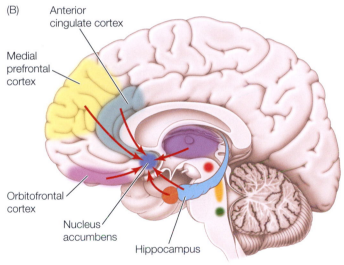

Figure 9.2   The dopaminergic innate reward system. The brain in (A) depicts axons that originate in the ventral tegmental area (red) and release dopamine in the nucleus accumbens and elsewhere. The locus coeruleus and dorsal raphe also modulate drug reward. The brain in (B) depicts regions involved in reward: the medial prefrontal cortex, the orbitofrontal cortex, the anterior cingulate cortex, the thalamus, the hippocampus, and the amygdala, all of which send excitatory axons to the nucleus accumbens. Addictive drugs alter this reward circuitry. (After Robison and Nestler 2011.)

negative emotional states are transiently eliminated by psychostimulants and opioid drugs, respectively. The two types of reward systems control the so-called appetitive and consumptive behaviors, which are based on feed-forward and negative feedback circuits of homeostasis.

The innate reward and pain-suppression systems

The dopaminergic innate reward system is centered on a transcription factor, ΔFosB, which is expressed in the nuclei of nerve cells in the ventral tegmental area, the nucleus accumbens, and part of the prefrontal cortex of the brain (Figure 9.2). The expression of this transcription factor is regulated by signals from the dopaminergic reward pathway. Many of the drugs that cause addiction, such as cocaine and methamphetamine, increase dopamine signaling and enhance ΔFosB expression, which then leads to a sequence of events that cause a subjective feeling of pleasure. It is thought that behavioral addictions, which include gambling and compulsive sex, also are mediated by this reward system (Olsen 2011).

The opioid reward system elicits the pleasure of satisfaction, whether it is caused by food consumption or pain relief. Thus, the pituitary gland can produce neuropeptides called endorphins that act like opioids, inhibiting the transmission of pain signals and producing a feeling of euphoria (Simantov and Snyder 1976). Interestingly, opioids can also be released in anticipation of possible pain—for example, as part of the fight-or-flight response—where they can help maintain function temporarily until it is safe to rest (or collapse).

Hijacking the reward system

Opiates, such as morphine and heroin, activate the opiate receptors that naturally mediate endorphin signaling. Cocaine inhibits dopamine reuptake transporters, thereby stimulating the dopaminergic reward system. Nicotine binds to acytylcholine receptors that trigger the release of dopamines in the ventral tegmentum and to nicotinic receptors in the adrenal medulla that trigger the release of adrenaline. Alcohol enhances the binding of glutamate, triggering the expression of ΔFosB, and the activity of nicotinic acetylcholine receptors, triggering the release of dopamines as well as having many other effects. Amphetamines are powerful stimulants of the central nervous system with many effects, one of which is to trigger enhanced expression of ΔFosB. Thus, all these compounds set in motion a cascade of events that lead to the sensation of pleasure, the cessation of pain, or both.

Self-medicating with plant compounds

Both primates and insects that are infected with pathogens are known to consume plants containing compounds with curative effects. The self-medication hypothesis, in contrast to the hijacking hypothesis (which sees addictive drugs as evolutionary novelties), posits that humans have long been exposed to plant compounds with curative properties, have sought them out when infected, and have evolved a reward system that motivates them to do so (Hagen et al. 2009). One study found a negative correlation between smoking and worm burden; the lower worm burden occurred in individuals with *CYP2A6* alleles that metabolized nicotine slowly, suggesting that nicotine is a worm poison (Roulette et al. 2014). This study, however, was conducted in a human population in Africa that could not have been exposed to nicotine for more than 300 years, which is not enough time for much evolution. Further work is needed.

Genetic variation in susceptibility to addiction

Addictive substances are often sensed and processed by the receptors and enzymes that process drugs, and we saw in Chapter 2 that there is extensive human genetic variation for capacity to process drugs. It is therefore no

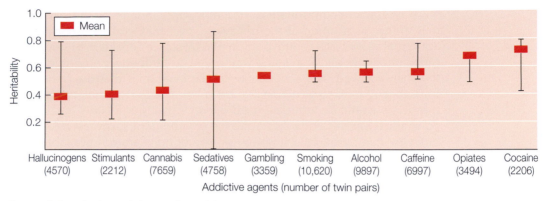

Figure 9.3   The heritabilities of 10 addictive disorders, means and ranges. (After Goldman, Oroszi, and Ducci 2005.)

surprise that humans vary genetically in their susceptibility to drug addiction, and the heritability of addictive disorders is indeed quite high, ranging from 0.4 to 0.7 in large twin studies (Figure 9.3).

Variation in nicotinic acytylcholine receptor genes is associated with modulation of the response of the reward system in healthy adolescents (Nees et al. 2013). Variation in the aldehyde dehydrogenase enzyme influences the aversive response to ethanol in East Asians, and several chromosomal regions have been found that influence the level of response to ethanol, which varies considerably among individuals (Enoch 2014). Similar comments apply to most addictive substances. The point is that recent evolutionary experiences have selected for genetic variation in the ability to sense and process ingested molecules, some of which have structures like those of addictive substances. As a result, people vary in their innate susceptibility to addiction.

◼ SUMMARY

Drug addiction results from ingestion of substances that trigger systems that evolved to shape adaptive behaviors through rewards or to suppress pain in emergencies. Such exposure may be a cultural novelty or an experience with an evolutionary history. In either case, it illustrates the principle that every adaptive system contains a vulnerability that can, under the right circumstances, permit the expression of a disease. That vulnerability often varies genetically among individuals.

Anxiety, depression, and obsessive-compulsive disorders

Most mental disorders are extreme expressions of behaviors that have normal adaptive counterparts. One can envision two very different reasons these extreme forms of behavior exist in human populations. First, as discussed at the beginning of this chapter, behaviors are products of many interacting components. Thus, at a population level, there must be tails in the distribution of the behavioral trait that would be considered abnormal as well as a central portion that would be considered normal and presumably beneficial in some way, at least in the environment in which the traits evolved. The second kind of explanation is that the behavioral traits that evolved to be adjustable are vulnerable to dysregulation, as with the homeostatic system discussed earlier. Thus, our anxiety level is not constant: it needs to be adjusted to the environment. Likewise, risk-taking behavior can be beneficial when risk-free options are not available. This adjustable nature of behaviors may make them vulnerable to dysregulation, leading to maladaptive behaviors that can be further enhanced by environmental mismatch and genetic predisposition.

Anxiety

Fear and panic are intense forms of anxiety that are clearly adaptive in appropriate circumstances: they help us maintain supportive social contacts, avoid danger, and escape in emergencies. Fear of snakes or spiders is usually helpful. We are terrestrial organisms with long exposure to them, and we do not have similar reactions to equally or more dangerous marine organisms like blue-spotted octopuses or deadly jellyfish. Some level of fear of strangers, perhaps no more than an initially cautious reserve, is probably also helpful given that some humans are among the most dangerous creatures on the planet. As another example, separation anxiety helps infants and young children maintain parental support. People who lack the capacity for fear and panic would probably not survive long in any challenging environment. As with many other defensive functions, an exaggerated and excessive feeling of anxiety can be so detrimental that it becomes pathological in the sense that it interferes with normal behaviors.

The smoke detector principle and the null hypothesis of complex systems

When a response is cheap and protects from serious injury, it pays to set the threshold to trigger the response at quite a low level. Although doing so can result in many false alarms, they do not cost much, and the low threshold ensures that almost all real threats trigger helpful responses (Nesse 2005). The biological systems that react to threats are built of many components, each of which could vary, resulting in a distribution of sensitivities in an entire population that looks like Figure 9.1; the tails of the distribution would then contain the pathological extremes that manifest as clinical phobias and other anxiety disorders. Although not contributing to the treatment of the individuals who inhabit the tails of the distribution, this insight may help explain why they exist at all.

Depression

The leading evolutionary hypothesis for mood disorders is that normal, functional variation in mood evolved as a mechanism to allocate effort in proportion to reward: "When payoffs are high, positive mood increases initiative and risk-taking. When risks are substantial or effort is likely to be wasted, low mood blocks investments" (Nesse 2015, in press). This idea fits easily into the framework advanced in Figure 9.1: normal variation near the mean is functional, but the extremes of mood, both positive and negative, are here seen as unavoidable population-level negative consequences of a response that is usually helpful.

Obsessive-compulsive disorder

OCD has substantial heritability and about 1% incidence; it "is character-ized by ritualistic repetitive behaviors and fears that some small oversight will lead to disaster" (Nesse 2015, in press). It may (cf. Figure 9.1) represent the maladaptive extreme of a response that is functional when closer to the mean: habits. Habits are useful because their function is proven by experi-ence and because we do not have to spend time thinking about them. The exaggeration of a habit, however, could become a destructive fixation; this idea is untested. Similarly, foraging and pathogen avoidance are adaptive behaviors, but their exaggerated forms, hoarding and germophobia, are pathological extremes of the behavioral spectrum.

■ **SUMMARY**

Many mental disorders have easily recognizable, normal behavioral counterparts, emotional states, and personality features. Anxiety, de-pression, and OCD are just a few examples of the pathological states that are extreme versions of adaptive behaviors. Mild versions of per-sonality disorders, including histrionic (attention-seeking), narcissistic, and dependence disorders, are also quite common and can be easily recognized in individuals in most social groups.

Autism and schizophrenia

Although evidence supports the idea that disruption of genomic imprinting can explain some of the variation in parental investment that is mediated by fetal growth and gestation length (see Chapter 7), not all the parent-of-origin imprinted genes are expressed in the fetus and in fetal tissue in the placenta. Some genes that are expressed in the brain and influence behav-ior also have parent-of-origin imprinting and vary in copy number (deletions and duplications). That variation in copy number produces syndromes that suggest the origins of vulnerabilities to autism and schizophrenia.

One gene whose copy-number variation has such consequences is found on chromosome 15. When the maternally inherited copy is deleted or disrupted, allowing paternal genes to be expressed unimpeded, the result is Angelman syndrome. Infants with Angelman syndrome demand to suckle very frequently, cry a lot, are hyperactive, sleep little, and have high risk (40%–80%) of autism. In contrast, when the paternally inherited copy is deleted or disrupted, allowing maternal genes to be expressed unimpeded, the result is Prader-Willi syndrome. Infants with Prader-Willi syndrome suckle poorly, cry weakly, are inactive and sleepy, and have high risk (30%–70%) of developing psychoses as adults.

Deletion or duplication of two other genetic regions results in similarly contrasting risks of autism and psychoses. Deletion of chromosomal region 7q11.2 results in Williams syndrome. Children with Williams syndrome have altered verbal skills and severe visual-spatial deficits; they are hypersocial, fascinated by faces, and highly empathetic; and they suffer high rates of anxiety and phobia disorders. Duplication of the same region results in children who have severe language delay, increased risk of seizures, and very high rates of autism and autistic features. Deletion of chromosomal region 7p11.2 results in Smith-Maginis syndrome. Children with Smith-Maginis syndrome have relatively strong verbal skills and high levels of sociability, but they also are at increased risk of bipolar and mood disorders. Duplication of the same region results in Potocki-Lupski syndrome. Children with Potocki-Lupski syndrome are at increased risk for seizures and autism.

Such observations led Crespi and Badcock (2008) to propose that autism and psychosis are diametrically opposed mental disorders that represent the extremes of a spectrum that is created by imbalances in maternally and paternally derived gene product in brain tissue. That balance can be shifted either by disrupting parent-of-origin imprinting or by variation in gene copy number (Figure 9.4). Maternal biases in gene expression increase the risk of psychoses; paternal biases increase the risk of autism.

Autism is defined by specific deficits in social reciprocity, language, restricted interests, and stereotypical behavior. In extreme cases, patients never learn to speak. Less extreme cases are described as autism spectrum

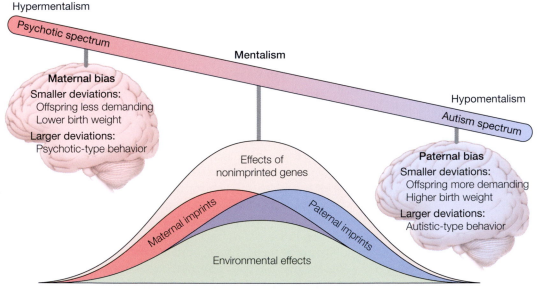

Figure 9.4   The Crespi-Badcock hypothesis for mental disorders. (After Byars, Stearns, and Boomsma 2014.)

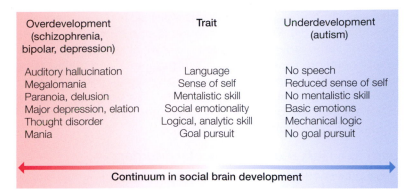

Figure 9.5 Autism and psychosis may represent the opposite ends of a spectrum of the development of the human social brain.

disorders. Psychosis is a term that covers a variety of mental disorders, including schizophrenia, which is characterized by hallucinations, delusions, paranoia, thought disorders, and disregulated mood; bipolar disorder, which is characterized by alternating mania and depression; and major depression. Mentalistic cognition refers to sensitivity to and awareness of what others may be thinking. All three psychotic disorders share hyper-mentalistic cognition, hallucinations, and delusions. The contrasts between autism and psychosis in a series of conditions led Crespi and Badcock (2008) to propose a continuum in social brain development affecting language, sense of self, mentalistic skill, social intelligence, analytic skill, and goal pursuit (Figure 9.5).

Meta-analysis using copy-number variation

By 2009 there had been 18 papers published on the association of autism or schizophrenia with deletions or duplications of chromosomal region 1q21.1, a region that contains unknown control elements and seven genes. One meta-analysis found a significant and striking contrast, with deletions in this region increasing the risk of schizophrenia and duplications increasing the risk of autism (Table 9.1). When they extended this approach to cover seven chromosomal regions, Crespi, Stead, and Elliot (2010) found four in

TABLE 9.1 Copy number variants affecting autism and schizophrenia in region 1q21.1

	Deletions	Duplications
Autism	2	10*
Schizophrenia	15*	4

Source: Crespi, Stead, and Elliot (2010), Table 1, p. 1737.

Note: Counts with asterisks are cases of statistically significant risk in case-control studies; $p = 0.001$.

which there was diametrically opposed reciprocal risk. In two of these four regions, deletions raised the risk of schizophrenia, and duplications raised the risk of autism. In the other two regions (one of which is 1q21.1), the risks were reversed: deletions raised the risk of schizophrenia, and duplications raised the risk of autism. In three additional chromosomal regions studied, there was no pattern associating the risk of mental disorders with deletions or duplications. Copy-number variation contributes to the risk of autism and schizophrenia, but it is not the whole story.

Size at birth as a marker of risk of mental disorders

The theory of parent-offspring conflict predicts, and experiments on mice confirm, that when the mother's interests are overexpressed, infants will be born smaller and earlier than average, and when the father's interests are overexpressed, they will be born larger and later than average. Such disturbances to normal genetic equilibrium may be caused by mutations that either disrupt imprinting patterns or cause copy-number variation, both of which can change the balance of maternally and paternally derived transcripts in the cells of the developing infant. We cannot genetically manipulate humans as we can mice, but we can investigate birth weights and gestation lengths in the rare cases in which mutations create contrasting syndromes and the balance of maternally and paternally derived transcripts is disturbed. Such contrasting pairs of syndromes include Beckwith-Wiedemann (paternal bias) versus Silver-Russell (maternal bias) and Angelman (paternal bias) versus Prader-Willi (maternal bias). An analysis of more than 1.75 million births in Denmark between 1977 and 2009 showed that infants with syndromes of paternal bias were significantly heavier and longer than average, and infants with syndromes of maternal bias were significantly lighter and shorter than average, adjusted for gestation period (Byars, Stearns, and Boomsma 2014).

If the imbalance in parental transcripts influencing size at birth continued after birth and was expressed in the brain, as suggested by the contrasting risks of autism and schizophrenia in those two pairs of syndromes, size at birth could serve as a marker of risk of mental disorders. In that Danish sample of more than 1.75 million births between 1977 and 2009, more than 95,000 children were diagnosed with an autistic or psychotic disorder (Byars, Stearns, and Boomsma 2014). When many other risk factors were controlled (including mental illness and age of parents and parity and sex of infants), the risk of autism increased and the risk of psychosis decreased with size at birth (Figure 9.6). The contrast in risks is less evident at the smallest birth sizes in very premature infants. There does appear to be a spectrum of reciprocal risk of mental disorders, as suggested by Crespi and Badcock (2008), and it is related to a marker of parental conflict over investment in offspring: birth size.

This connection between the biology of pregnancy and birth and the risk of mental disorders is striking, unexpected, and would never have been investigated had it not been suggested by the evolutionary theory of

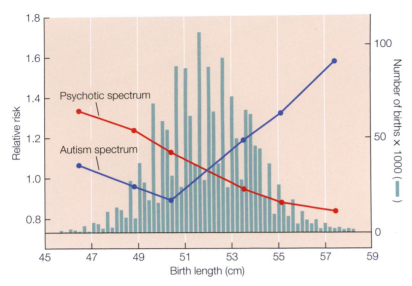

Figure 9.6   Risk of autism and psychosis in 1.75 million infants born in Denmark between 1977 and 1990 is related to size at birth, a marker of parental conflict over investment in offspring. (After Byars, Stearns, and Boomsma 2014.)

genomic conflict. It is, however, important to emphasize that the pattern in Figure 9.6 is consistent with Crespi and Badcock's interpretation. It does not demonstrate it, however, because it is a correlational pattern, not the result of an experiment, and the mechanisms causing the risk were not available for investigation. It is also important to note that autism and psychosis appear to have many causes, and the one highlighted here probably does not account for more than 10%–20% of the variation in risk.

■ SUMMARY

The connection between genomic conflict and mental disease is made through disturbances to a balance of genetic transcripts influencing behavior that usually results in healthy minds and normal behavior. Here mental disease is not thought to be selected; rather, it is thought to be the pathological by-product of the disruption of an equilibrium that usually results in healthy children. This interpretation of the causes of autism and psychosis is supported by a growing body of several types of evidence, but other interpretations have not been not ruled out and are not mutually exclusive.

Individual Health versus Population Health

This chapter discusses the tensions that arise from the conflicts between individual health and population health. Physicians usually focus on individual patients, whereas public health officials and evolutionary biologists usually focus on populations. When physicians treat individuals, their decisions have population-level consequences for epidemiology, demography, and natural selection, a striking case of biological evolution driven by cultural intervention. The chain of causation between individual medical decisions and long-term, large-scale, population-level consequences is not always noticed and deserves more attention. It is a connection with which evolutionary biologists are familiar.

Some of the cases we describe also exemplify another general issue, the conflict between the welfare of the individual and the welfare of the group, a conflict that often involves externalization of private costs and exploitation of public goods by private interests. These cases reflect decisions made by patients and doctors within frameworks of incentives created by political and commercial institutions. Although these terms from economics and political science—welfare, private costs, public goods—imply monetary costs and benefits, money does not capture all their important aspects; in medicine and public health, there are moral dimensions and social values that transcend strictly economic calculations, however much they may be constrained by them.

The scientific and moral landscape

The debate in the United States over how health care should be financed provides a useful introduction to the subject of individual health versus population health because it contains many of the conflicts and much of the complexity of what follows. Single-payer plans in which the government insures everyone and everyone is required to participate provide the greatest coverage at the lowest cost and thus the greatest benefit to the entire population, but they do not give individuals the freedom to decide whether they want to be insured at all, and if so, by whom. The young and the healthy have less incentive to participate than do the old and the sick, but only if everyone participates can the greatest reductions in cost be realized. From the point of view of the entire population, a young, healthy person who chooses not to pay into the insurance pool is a cheater parasitizing a public good. That person's decision may be well justified in terms of short-term self-interest, but from the point of view of the collective, it is a cost to the group because thousands of such individual decisions accumulate to raise enormously the costs of health care that must eventually be borne by all. The decision of how health care should be financed is thus influenced by deeper, more general assumptions about the relative value of individual freedom versus social welfare: it is a decision that probes our values.

Here we explore this intersection of the scientific and moral landscapes that, in summary, holds three important lessons.

■ SUMMARY

1. The conflict between individual interest and group interest is real and is not going away. It can be mitigated, but it cannot be eliminated.
2. Medical practice and the management of public health have changed the human environment so radically that they are now major drivers of natural selection on traits of medical and demographic importance.
3. Sometimes when medicine solves old problems, it creates new ones.

Population consequences of medical decisions

We begin with two places where individual medical decisions have significant population consequences. The first explores herd immunity as a public good and reminds us that widespread use of imperfect vaccines should cause the pathogens targeted to evolve increased intrinsic virulence (see Chapter 5).

The second looks at the population consequences of individual medical decisions as exemplified in the conundrums posed by antibiotic use. When an individual patient is vaccinated or treated for an infection with antibiotics, that individual benefits, but these interventions can create significant indirect costs at the population level. They do so in different ways.

Then, third, we turn to how medical practice and information about health have been contributing to sweeping changes in the causes of mortality, the demographic structure of populations, and the traits on which natural selection is acting.

Vaccination

The great vaccines of the twentieth century—the vaccines that protect against measles, mumps, rubella, pertussis, diphtheria, smallpox, tuberculosis, and polio—are all sterilizing. When they succeed in their stimulation of the immune system, they protect completely, often for life, removing the vaccinated individual from the pool of susceptible people who can transmit the disease. When the proportion of immune individuals rises to a high enough level (about 90% for measles), the disease can no longer spread through the population, which is then said to be protected by *herd immunity* (R. Anderson and May 1985).

Herd immunity is a public good created by those who decide to vaccinate that protects those who have not yet been vaccinated. It creates a situation in which a small minority of people can refuse to be vaccinated and get away with it as long as not too many do so. However, when many refuse vaccination—as recently happened with vaccines for measles, mumps, rubella, diphtheria, pertussis, and tetanus—diseases that had nearly disappeared can again break out. Recent outbreaks of measles and pertussis cluster strongly with the geographic distribution of parents who refuse to have their children vaccinated (Omer et al. 2009). Those who suffer the most are infants younger than 6 to 12 months old, who are too young to be vaccinated. For them, the diseases can be deadly. For example, in the California pertussis epidemic of 2010, all deaths occurred in infants less than 3 months old (Winter et al. 2012). The refusal to vaccinate raises an important moral issue about indirect consequences.

Many parents refused vaccination for their children because they were afraid that vaccination would significantly increase the risk of autism. Andrew Wakefield, a British physician, promulgated that idea in a 1998 paper shown to be fraudulent in 2010, when the *Lancet* retracted Wakefield's publication. Wakefield was struck from the Medical Register and barred from practicing medicine in the United Kingdom. Considerable damage had already been done, however. Immunization rates in Britain dropped from 92% to 73% and were as low as 50% in some parts of London. The Centers for Disease Control and Prevention estimated that more than 125,000 children in the United States did not get the vaccine for measles, mumps, and rubella. Celebrities picked up on Wakefield's idea and spread it in the media, and television producers showcased them in popular television programs. Wakefield was found guilty of fraud and banned from medical practice. Was he guilty of murder? Were celebrities and television producers guilty accomplices in a crime that continues to kill innocent children? Or did they just make an innocent mistake with unfortunate collateral damage? Whatever your conclusion, one thing is clear: individual decisions about vaccination have had very serious population-level consequences that continue to play out.

Vaccines are involved in another individual-versus-population tension. As we saw in Chapter 5, imperfect vaccines—vaccines that do not produce sterilizing immunity—have important consequences for pathogen evolution. The widely used vaccine that protects against human papillomavirus, the causative agent of cervical cancer, is an imperfect vaccine that is effective for only two to four of the many circulating strains of this virus. All candidate malaria vaccines under development are imperfect; they will protect only some of the people vaccinated. The widespread use of imperfect vaccines can drive the evolution of increased virulence in the target pathogen because by allowing the patient to live longer, the virulent strains have more time to transmit, thus altering the balance in the virulence-transmission trade-off to favor the evolution of greater intrinsic virulence in the pathogen. Even an imperfect vaccine benefits the vaccinated individual by reducing risk of disease. There is, however, an indirect cost of the imperfect vaccine that is externalized to the level of the population. Individuals who are protected by the vaccine will not bear that cost, but individuals in whom the vaccine is ineffective as well as individuals who have not been vaccinated will be exposed to the risk of increasingly virulent pathogens. Again, decisions driven by individual cost-benefit analysis have population-level consequences.

Antibiotic therapy

When a physician decides to treat a bacterial infection, the patient gets all the benefit and some of the cost, but most of the cost is externalized and diffuse. If the therapy is successful, the patient gets well. In many cases, the difference is between life and death. The patient does bear some cost, but it is usually small. It includes the risk of harboring a resistant strain that cannot be treated in a future infection and a disruption of the microbiome that could increase risk of atopies, autoimmune disease, obesity, and virulent diarrhea. The diffuse, externalized cost is the contribution made by this individual decision to the global spread of antibiotic resistance, to which millions of such individual decisions have contributed. A similar chain of causation is activated when a farmer decides to treat a flock or herd of animals with antibiotics to promote rapid growth. The farmer receives a direct economic benefit, and the costs of resistance are for the most part externalized.

As we saw in Chapter 5, those externalized costs have now accumulated to pose deadly threats to millions of patients; the global spread of resistant strains has become a serious problem that threatens our ability to treat infections and do operations. These resistant strains arise and spread because individual decisions are driven predominantly by short-term individual cost-benefit calculations not often informed by population-level consequences. The individual benefits are immediate, local, large, and easy to understand. The population-level costs are delayed, diffuse, and, for any individual decision, small; they are harder to perceive and understand. These many small costs, however, add up to create a very large problem with serious consequences, and the health of everyone on the planet is directly or indirectly threatened by the spread of antibiotic resistance.

Weighing the good of the individual against the good of the population

The question of how to weigh the good of the individual against the good of the population involves more than differences between patients, physicians, and public health officers about vaccines and antibiotics. It also lies at the heart of political conflicts over health insurance, taxation, gun control, and much else. These conflicts all involve the tension between personal freedom and social responsibility, the focus of a debate that has been going on for centuries and will not be resolved anytime soon. The medical examples may be among the cases in which consensus can be more easily achieved because they show how decisions based on the freedom of the individual can have unintended consequences at the population level that put individual lives at risk. Consensus on such matters can only be reached in democracies if most of the people in a population understand how the problems are caused. Much remains to be done in educating the public about science, medicine, and the conundrums of medical policy.

■ SUMMARY

Private decisions about medical treatment can accumulate to have serious public consequences. Many of these private decisions are examples of bounded rationality: decisions that seem rational to the individuals who make them but that have public consequences that are ultimately irrational at both the public and the individual levels. Vaccination benefits individuals, but if not all are vaccinated, yet enough are vaccinated to create herd immunity, a perverse incentive is created for the remaining unvaccinated individuals not to get vaccinated because they can benefit from the protection of vaccination without accepting any of its risks. If enough decide not to be vaccinated, an opportunity for the target disease to reemerge is created. If the disease does reemerge, the youngest suffer the most. The use of imperfect vaccines creates another problem. Although protecting many people, such vaccines also drive the evolution of increased virulence, indirectly creating greater risks for those not protected by the vaccine. Antibiotic treatment benefits individuals but drives the rapid evolution of antibiotic resistance, which is now a global problem threatening billions.

The Great Transition

A great event, the largest change in human populations since the invention of agriculture, has been developing over the last 300 years, starting in France in the eighteenth century, transiting England in the nineteenth century, coming to completion in the United States in the twentieth century, and hitting midstride in parts of the developing world in the early twenty-first century. This complex of processes is variously referred to as the industrial revolution, the demographic transition, and the second epidemiological transition. Labeling it the industrial revolution emphasizes changes in labor, technology, and economics; labeling it the demographic transition emphasizes changes in birth and death rates, age distributions, nutrition, and growth; and labeling it the second epidemiological transition emphasizes changes in the prevalence of diseases, roughly from the category 2 diseases (mostly infectious diseases and malnutrition) to the category 5 diseases (cancer, chronic diseases, and degenerative diseases) discussed in Chapter 3. Because these processes share intertwined causes, we use the term *Great Transition* to cover them all. It corresponds to the later stages of the longer transition from medieval to modernity.

The Great Transition began with technologies made possible by the growth of science. Some technologies improved the productivity of agriculture, making possible the move from agrarian to urban societies, increasing the number of calories consumed per person per day, and improving both growth and health. Some technologies improved public hygiene: modern sewage systems and clean water supplies decreased the transmission of waterborne diseases, and especially important were the reductions in infant mortality caused by diarrhea. Some technologies gave physicians new tools: vaccines to prevent many infectious diseases (some ancient, some acquired with the agricultural revolution); sterilization to reduce infections in hospital wards and operating theaters; and antibiotics to kill microbial pathogens, whose causal role in many diseases was only recognized in the late nineteenth century. Urbanization reduced the breeding habitats of some disease vectors; population growth and habitat destruction exposed us to new, emerging diseases; global travel enabled long-distance transmission of vectors, diseases, and resistance genes; and massive use of fossil fuels drove global warming and climate change that are shifting disease and crop distributions today.

In short, processes that have not yet gone to completion are massively rearranging the human ecology of the planet. These effects are the cumulative products of billions of decisions, most of them based on short-term individual interest. Largely unaware of our collective power, we have been remaking the planet in a fit of inattention from which we are slowly waking up. The consequences go well beyond the scope of this book. Changes that are driven in significant part by decisions made by physicians, patients, and public health workers are discussed next (adapted from Corbett et al. 2015).

How declines in mortality and fertility are changing selection

Mortality has dominated the historical rise and fall of populations (Omran 2005), with rates ranging in premodern populations from a minimum of 30 deaths per thousand per annum to many times more during years of famine and epidemics. Fertility in these populations was 30–50 births per thousand per annum and was constrained by biological limits on interbirth intervals, fecundability, female survival to maturity, and marriage and contraceptive

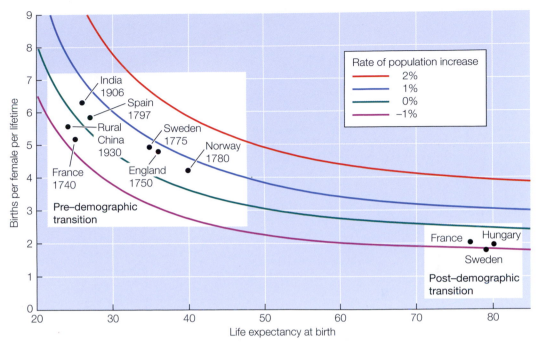

Figure 10.1   Shifts in key demographic parameters in selected countries before and after the demographic transition. Points represent mean fertility rates and expected life spans for countries before (left) and after (ca. 1980, right) the demographic transition. Curves indicate the rate of population increase. (After Coale 1986.)

practice (Omran 2005). Therefore, growth rates of populations were often positive, but they fluctuated considerably before the demographic transition. Following the start of the decrease in mortality (it continued to less than 10 per thousand per annum in the post–demographic transition populations), fertility began to decline in France in the decades following 1789 and in the rest of Europe between 1890 and 1920 (Figure 10.1). By 2003, 60 countries with 43% of the world's population had fertility at or below the replacement level of 2.1 children per woman (Lee 2003a). The pace of fertility decline is influenced by the perceived costs and value of children (Lee 2003a), life expectancy and literacy, and other cultural factors, including who controls female reproduction (Caldwell 1999).

The major changes in fertility and mortality that accompany the demographic transition shift the potential contributions of variation in mortality and fertility to the intensity of natural selection. Crow (1958) proposed an index of the opportunity for selection, *I*, which measures the potential for selection. When selection is directional and the trait is completely heritable and perfectly correlated with fitness, this index gives the standardized change in the mean of the trait after one generation of selection. When those conditions do not hold, it gives the upper limit to the selection that could occur.

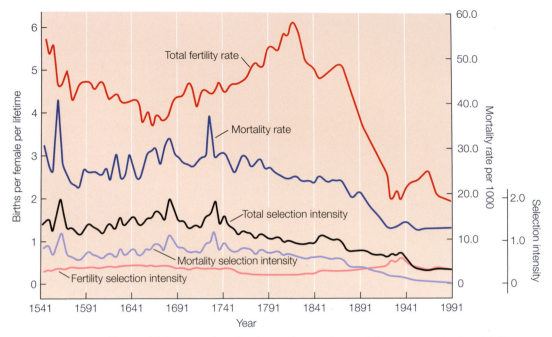

Figure 10.2   Trends in fertility and mortality and the opportunity for viability, fertility, and total selection in England, 1541–1991. Total selection intensity is the sum of mortality and fertility selection intensity. Dark red, total fertility rate; light red, fertility selection intensity; dark blue, mortality rate; light blue, mortality selection intensity; black, total selection intensity. (After Corbett et al. 2015.)

In countries with an extended record of fertility and mortality, such as England and France, the total selection intensity declines, and the relative intensity of fertility selection increases (Figure 10.2), suggesting that differences in fertility among individuals have recently started to exert stronger selective pressures than have differences driven by mortality. Analysis at the level of nations does not capture the considerable heterogeneity in the potential for selection that exists at a regional level. A study of a multigenerational cohort in Utah (Moorad 2013) confirmed that both the opportunity for selection derived from fertility and the total opportunity for selection increased during the demographic transition. In the more recent demographic transition in Gambia, the opportunity for selection derived from variation in fertility increased, but the total opportunity for selection decreased as mean fertility declined (Courtiol et al. 2013).

The effects of changed selection on life history evolution

Phenotypic evolution, the change in the mean values of traits between generations, is driven by natural selection; that is, it is driven by the association of heritable traits with differences in fitness, often measured as lifetime reproductive success. Differences in fitness only lead to evolution if the traits with which fitness is associated are inherited. Therefore, phenotypic models describing natural selection include three components: differences in fitness, heritability of the trait, and the amount of variation in the trait (Stearns et al. 2010). The response of one trait to selection can only occur if its heritability is not constrained by phenotypic and genetic correlations with other traits. Prediction of the magnitude and direction of phenotypic evolution must take these correlations into account (Stearns et al. 2010).

Evidence of natural selection has been reported in 15 longitudinal studies in pre- and postindustrial populations, all in what is now the developed world. These studies found that both women and men experienced phenotypic selection for earlier age at first birth; women experienced selection for greater age at last birth, and at menopause, greater weight, shorter height, and lower levels of cholesterol and systolic blood pressure (Stearns et al. 2010; Milot et al. 2011).

In other populations, results differed. For example, the demographic transition that occurred in several villages in Gambia between 1956 and 2010 resulted in some contrasting patterns (Courtiol et al. 2013). Prior to 1974, selection operated to decrease height and increase body mass index; from 1975 onward, selection operated to increase height and decrease body mass index.

Selection is not the only evolutionary component influenced by demographic transition. A study of Danish twin cohorts, which spans the entire Danish fertility transition, was able to track the contributions of genetic and environmental effects to fertility over this period. The authors found that genetic influences on early fertility increased strongly, accounting for up to 50% of the variability of early fertility in the most recent cohorts (Kohler, Rodgers, and Christensen 2002).

Overall, these studies suggest that the combination of earlier age at first birth and later age at last birth is broadening the temporal window of reproductive opportunity in the developed world. This finding can be interpreted as a response to a shift in the balance of selection components from selection driven predominantly by mortality to selection driven predominantly by fertility. Such a shift in selection, combined with a potential boost in heritability as environments become more standardized, is likely to significantly affect the evolution of fertility-related traits.

Increased longevity and the exposure of antagonistic pleiotropy

As a growing number of populations reach the last stage of the epidemiologic transition, cancers, degenerative diseases, and cardiovascular diseases have already displaced infectious disease as the major causes of death. Degenerative diseases and cancers were four of the top five causes of death globally in 2010, with major increases since 1990 in the mortality rates for ischemic heart disease (+35%), stroke (+26%), lung cancer (+48%), and diabetes (+98%) (Lozano et al. 2012). The large differences in the prevalence of these diseases (all category 5 diseases caused by declines in somatic maintenance) reflect differences among countries in the progress of demographic transitions. Infectious diseases still account for 55% of deaths and 68% of years of life lost (i.e., the aggregate difference between date of death and average life expectancy) in low-income countries compared with 14% of deaths and 8% of years of life lost in developed countries (World Health Organization 2013). If the transitions now under way proceed as expected, the contribution of degenerative diseases to total deaths should rise by 12%–45% for middle- and low-income countries between 2008 and 2030. Understanding the causes and consequences of noncommunicable diseases is therefore very important.

Because people are living longer, more of them are expressing genetic costs that previously were not often paid because people died earlier for other reasons. Evidence is mounting that some of the global burden of category 5 disease is contributed by the costs expressed later in life by antagonistically pleiotropic genes that contribute benefits earlier in life. Several studies, including genetic analyses of cancer risk and dementia, suggest that advantages in fertility or juvenile survival are linked to three genes that now confer risk at old age.

The first case concerns female reproductive cancers. Germline mutations in the *BRCA1* and *BRCA2* genes in populations in developed countries, although rare, account for between 1%–13% of ovarian cancer and 1%–5% of female breast cancer (Risch et al. 2006), but they are also associated with increases in lifetime reproductive success. In populations prior to the Great Transition, higher fertility may have mitigated the carcinogenic potential of these mutations because multiple pregnancies protect against reproductive cancer. Conversely, lower fertility in post–Great Transition societies would increase the risk of reproductive cancers, thereby weakening selection for these mutations. Genetic counseling may be promoting selection against these genes.

The second case concerns a relation between reproductive performance and the risk of cancer in general. The tumor-suppressor protein p53 plays a pivotal role in coordinating cellular responses to DNA damage. A common arginine/proline substitution in codon 72 (Arg72Pro) of the p53 protein is a polymorphism with differential effects on both longevity and fertility (Risch et al. 2006; Kang et al. 2009). Pro/Pro homozygotes and Arg/Pro heterozygotes have better survival after a cancer diagnosis than Arg/Arg homozygotes and no overall increase in cancer risk (Risch et al. 2006). However, the p53 Pro allele, through its regulation of leukemia inhibitory factor, is associated with blastocyst implantation failure and infertility, particularly in women younger than 35 years of age. Therefore, reduced fertility and delayed childbearing, both consequences of the demographic transition, may augment and diminish, respectively, positive selection for the major p53 Arg allele. That this allele experiences positive selection in Caucasian and Asian populations suggests that the change in the fertility rate is the main driver of this evolutionary change.

The third case concerns a relation between infant development and neurodegenerative and cardiovascular disease. The *APOE4* allele protects the cognitive development of children with heavy diarrhea but at the cost of greater risk of Alzheimer's disease and atherosclerosis later in life (Oria et al. 2005). Such an allele may therefore have been under strong positive selection in the past, may help to explain why Alzheimer's disease occurs frequently in later life, and may be selected against when health conditions early in life improve across the Great Transition.

These examples illustrate how prior to demographic transition, genes with larger individual fitness benefits (for fertility and infant survival), and smaller individual fitness costs (later-acting cardiovascular disease, cancer, and Alzheimer's disease) may have been subject to positive selection. Few of the people who carried them then survived long enough to experience the increased risk of cancer, cardiovascular, and degenerative disease because many died from infectious disease and in childbirth. Now those who carry the genes experience both larger costs due to longer life spans and potentially smaller benefits as the modern environment buffers the positive effect of genetic variation on fertility. The growing burden of category 5 diseases is thus caused in part by the Great Transition's uncovering of previously evolved and previously hidden costs.

■ SUMMARY

The Great Transition reduced mortality and fertility rates and continues to shift the direction of selection on major human life history traits, in particular those directly associated with fertility. In post–Great Transition populations, age at first birth is under selection to decrease, and age at last birth is under selection to increase. The window of reproductive opportunity is thus broadening. At the same time, more people are surviving to greater ages. This demographic shift is uncovering the costs implicit in genes previously selected for their benefits early in life that have pleiotropic costs expressed later in life. Some of those costs include increased risk of cancer and dementia. The Great Transition is thus simultaneously changing selection in contemporary populations and revealing evolutionary costs previously accumulated.

Nutrition, energetics, and fertility

Before the Great Transition, the standard of living of unskilled workers and farmers in the poorer areas of Europe, the Ottoman Empire, Egypt, China, India, and Japan did not differ much (Clarke 2008; Allen 2014). The average person of 1800 was not much better off than the average person of 100,000 BCE (Clarke 2008; Floud et al. 2011). Most pre–Great Transition populations were undernourished: short stature was widespread in agrarian and pastoral populations (McKeown 1976, 1983). Seasonal food shortages and chronic undernutrition raised mortality and, if prolonged, caused crisis mortality and reductions in the fertility rate. Births were often strongly seasonal, with a drop in conceptions in the hungry season prior to harvest or before monsoonal rains (Panter-Brick 1996). Workers in subsistence agriculture had caloric intakes from 1500 to 2500 kcal/day. Hunter-gatherers and nomadic pastoralists fared little better, although there were some pockets of abundance in which the work required to obtain sufficient food consumed only 2–5 hours per day (Clarke 2008; Floud et al. 2011).

During the Great Transition, compounding improvements in agriculture, transport, and food distribution uncoupled cycles of mortality and fertility from the price of grain (Wrigley and Schofield 1981; Fogel 2004). During the nineteenth century and into the mid-twentieth, improvements were slow and suffered some reversals because of rapid urbanization (Szreter 2004), but between 1961 and 2003, there was a 25% global increase in food supply, or 550 kcal per person. Although less exposure to pathogens through piped drinking water, sewerage infrastructure, and vaccinations did help reduce mortality, much of that reduction was caused by improved nutrition and the accompanying increase in immunity to infection (Di Rienzo 2006). Fertility rates can increase in the early phases of demographic transition (Dyson and Murphy 1985) due in part to improvements in nutrition, but they then reverse due to conscious limitation of family size.

The wide variations in fertility seen in natural fertility populations were caused by genetic, social, and environmental factors. In descending order of importance they were the duration of lactational amenorrhea; the proportion of women married at each age; the behavioral, nutritional, and genetic factors that affect time to conception; age at menarche; pathological sterility; age at menopause; fetal loss; and the length of gestation. All these factors can be influenced by energy availability. If the more distal determinants of fertility, such as religious affiliation or wealth, influenced fertility, they must have done so by affecting one of the more proximate determinants (K. Campbell and Wood 1988).

The different components of female reproduction vary considerably in their sensitivity to the nutritional and work environments. Human ovarian function is quite sensitive to energy balance and energy flux. The duration of lactational amenorrhea, gestation, and birth weight are somewhat less sensitive to metabolic load and maternal energy availability than are other components. Reproductive effort during gestation and maternal milk production (Ellison 2003) are quite insensitive: they are relatively buffered from energy constraints. The evolved rule of thumb appears to be that if stressed, do not get pregnant, but if both stressed and pregnant, invest enough to have a surviving infant and then lactate enough to keep it going.

In societies since the Great Transition, fertility levels fall to near or below replacement levels, even below the number of desired children (Bongaarts 2001). Lactational and nutritional amenorrhea recede as important determinants of fertility. Age at menarche falls by approximately 5 years, from 17–18 years old in preindustrial populations to 12–13 years old in postindustrial populations, and age at menopause rises from 42–43 years old to 50–51 years old (Wood 1994), extending by almost a decade the potential reproductive life span. The postponement of childbearing is a striking feature of post–Great Transition fertility schedules, as is the importance of early fertility (Kohler, Rodgers, and Christensen 2002). Many other factors also contribute to determining contemporary fertility, including involuntary infertility through delayed marriage or divorce, the instability and nature of some sexual unions, changes to the sexual division of labor and the costs of raising a child, and effective contraceptive technologies (Balbo, Billari, and Mills 2013).

Metabolic adaptations that arose in our lineage buffer both reproduction and the brain from ecological variation (Holliday 1989; Kirkwood and Shanley 2005). They also optimize the allocation of resources to reproductive effort over the life span (Wade and Jones 2004). To optimize lifetime reproductive success, individuals are always obliged to allocate food energy among three major functions: basal metabolism and respiration, immune function and bodily maintenance, and growth and reproduction. The dramatic environmental and energetic changes of the Great Transition required a reallocation of available food energy that was implemented by both phenotypic change and genotypic change. For example, phenotypic or genetic metabolic factors that helped sustain ovulation or shorten the duration of lactational amenorrhea under conditions of seasonal food shortage would then have increased fertility, but this fertility advantage of these phenotypes and genotypes was probably lost or even reversed with the elimination of food shortages.

The shift from infectious to degenerative disease

Global patterns of infectious disease have shaped human genetic diversity (see Chapter 2) both by influencing the distribution of alleles of resistance genes (e.g., for malaria) and by shaping the association of high fertility (Guégan et al. 2001) and birth weight (Thomas et al. 2004) with increased pathogen densities. Pathogen pressure is generally highest in tropical regions and recedes with increasing latitude (Guégan, Prugnolle, and Thomas 2009). Beginning in the sixteenth century, European colonial expansion initiated a major shift in human-microbe relationships. The transoceanic dispersal of diseases such as smallpox and measles to naive populations in the Americas and Australasia had devastating effects on adult and child mortality, as did syphilis in the reverse direction on similarly naive populations in Europe (McNeill 1976).

Initially, the rapid urbanization that accompanied the Great Transition multiplied opportunities for disease transmission and, for much of the nineteenth century, stalled the decline in mortality in England, which had begun in the eighteenth century (Darwin 2002). Mortality from enteric and waterborne diseases such as cholera and typhoid then fell sharply in response to the introduction of sewers, piped water, and pasteurized milk. Reductions in mortality from diseases transmitted by food, vectors, and personal contact were more difficult to achieve; they required new technologies, including refrigeration and antisepsis, and improvements in health literacy, health promotion, and regulation (McKeown 1983; Szreter 2004). Although mortality from airborne diseases such as tuberculosis, smallpox, diphtheria, and whooping cough did decline with improvements in the built environment and living conditions, major reductions in the mortality caused by infectious disease were only achieved with the development and use of effective childhood vaccines and antibiotics. Thus, changes in pathogen pressure are very recent and highly variable across the globe.

The drop in infectious diseases decreased selection for immune defenses, but evolutionary shifts in allocation to defense will take many generations. Meanwhile, the post–Great Transition environment has revealed vulnerabilities in an immune system that had previously been highly beneficial (see Chapter 8).

■ SUMMARY

Improvements in the quality and quantity of food, improvements in the reliability of the food supply, and reductions in childhood mortality from infectious diseases are causing sharp decreases in selection on genes, influencing metabolic resource allocation and disease resistance that have long shaped human phenotypes. The rise in atopies and autoimmune disease appears to be causally linked to the sweeping changes in our microbiomes that have accompanied this major cultural shift.

Recent changes in demography and selection have revealed previously hidden costs of reproduction that are now expressed late in life in the form of increased genetic risk of degenerative diseases. They have also revealed added conflicts between individual and group interests. Navigating such issues in which biological evolution and cultural evolution interact to change health and disease requires facility in moving back and forth, from the interests of individuals to the interests of populations. Thinking about evolutionary processes helps develop that facility.

Open Questions and Other Issues

In this final chapter, we discuss the questions that call for solutions through additional research, explain why we have not discussed in more detail some issues often identified with evolutionary medicine, and conclude by summarizing the differences between classical and evolutionary medicine. We draw considerably on material from a recent review article (Stearns 2012).

Open questions

The evolutionary perspective on important medical issues raises questions about points where research, both basic and translational, can improve the science that supports our efforts to reduce suffering and save lives. Here we have selected for discussion those we find most important. We invite you to extend the list.

Can we develop evolution-proof antimicrobial therapies?

Because resistance rapidly evolves every time a new drug or insecticide is used, we are losing the race to produce drugs to treat bacterial infections and insecticides to kill insect vectors. Evolutionary biologists are therefore trying to discover and test evolution-proof interventions that will not cause the evolution of resistance or will, at least, cause resistance to evolve less quickly than it currently does (see Chapter 5).

Phage therapy—killing pathogenic bacteria in humans with viruses that evolved to infect bacteria—is one option. By multiplying, phage increase in proportion to bacterial density, and they can coevolve on the same timescale as bacterial defenses. Phage therapy was first used in 1926, but it was then neglected in the West; it continued to be used in Poland and Russia, however, where more than 1000 patients have been treated (Levin and Bull 2004). Because of the dramatic recent increases in deaths caused by multiply-resistant bacteria, interest in phage therapy has intensified (Reardon 2014), with a focus on practical applications and potential side effects (Chibeu et al. 2012; Tsonos et al. 2014; Viertel, Ritter, and Horz 2014). It offers significant possibilities.

A second option is disrupting bacterial public goods, which include the siderophores that they secrete to scavenge iron. Treatments with gallium that quench siderophore activity destroyed the siderophores' usefulness and attenuated virulence and bacterial growth. In contrast to conventional antibiotic therapy, the siderophores did not lose efficacy over time; it is difficult for a rare mutant producing a public good to invade a population whose residents are not producing that good while benefiting from it (Ross-Gillespie et al. 2014). Another public good consists of the signals that bacteria use to coordinate their attack (West et al. 2007; S. P. Brown et al. 2009). Bacteria monitor their local abundance with a quorum-sensing system. They do not attempt an infection or other collective behaviors, including biofilm formation, until enough are present to overwhelm host defenses. Disrupting such signals is another possibility, but it has not yet been as convincingly tested as siderophore quenching.

In the case of insect vectors, it may be possible to exploit vulnerabilities in their life cycles. Selection on the young is usually more intense than selection on the old, so if insecticides that kill mosquitoes only several days after exposure are designed, they will elicit the evolution of resistance more slowly than quick-acting insecticides. To make vector control sustainable, late-acting pesticides could be combined with larvicides that reduce life span and decrease biting in adults (Koella and Lorenz 2009; Koella, Lynch, and Thomas 2009). Where the costs of resistance are significant, this approach should be evolution-proof in the long term, solving the problem of mosquito resistance for many years (Read, Lynch, and Thomas 2009).

Can we switch the host-pathogen interaction from resistance to tolerance?

Because physiological and biochemical functions are buffered at many levels, it is possible to tolerate the malfunction of part of a redundant process. For example, African green monkeys and sooty mangabeys tolerate infection by SIV because even though their immune cells are infected with the virus, just as human cells are with HIV, the infection does not produce debilitating pathology: some of the reactions that occur in humans do not occur in the monkeys (Pandrea et al. 2008; Jacquelin et al. 2009). We need to understand the conditions under which it pays to tolerate an infection rather than resist it, the trade-offs in which tolerance is involved so as to judge when to promote it, and the mechanisms that mediate the transition from resistance/virulence to tolerance/commensalism. What does inflammation trade off with, and why? Can endothelial damage and atherosclerosis be reduced by strategies that seek to modulate such trade-offs?

Does virulence evolve when leaky vaccines are used?

The human intervention with the greatest potential to make pathogens more lethal—vaccination with nonsterilizing (leaky) vaccines—has been experimentally confirmed in mice and chickens (Gandon et al. 2001; Mackinnon and Read 2004; Mackinnon, Gandon, and Read 2008). Better understanding is urgently needed; use of the human papillomavirus (HPV) vaccine is spreading (Teo and Locarnini 2010), and any malaria vaccine will produce large-scale evolutionary experiments involving human subjects. We need to know whether increased virulence will evolve, and, if so, how fast and how far that will happen. The virulence of HPV must be monitored, and monitoring programs should be designed for malaria virulence before any vaccine is released. How to manage this type of virulence evolution should be investigated, and treatments for the more virulent strains that are anticipated to emerge should be developed.

Can we treat cancer by slowing somatic evolution?

The evolution of the resistance of cancers to chemotherapy strongly resembles the evolution of bacterial resistance to antibiotics. In both, strong selection caused by large doses applied for long periods rapidly selects for resistance. Many bacterial resistance genes evolved long ago and far away for other purposes and are horizontally transferred, fully developed and ready for use, into pathogenic populations. Many characteristics of cancer cells are present in stem cells, where they evolved for other purposes and are ready to emerge when mutations shift the control of their expression. Research into the management of resistance evolution should seek to combine insights from microbiology and oncology.

Such research has recently begun. One result deserves emphasis: aggressive chemotherapy may defeat its purpose by efficiently selecting for resistant cancer clones that shorten patient life span (Gatenby et al. 2009; Silva and Gatenby 2010; Read, Day, and Huijben 2011; Gillies, Verduzco, and Gatenby 2012; Silva et al. 2012). The recommendation only to use as much chemotherapy as is needed to keep the growth of the cancer under control is not one that either doctors or patients will readily accept. Additional demonstration of the effect in clinical trials is urgently needed.

What are the ultimate reasons for susceptibility to cancer?

Do we have naturally evolved defenses against cancer, and, if so, how do they work? We know that the immune system may be involved, and targeted interventions using immune therapy have succeeded in some cases (Z.-X. Wang et al. 2014) but not in others (Rogers, Veeramani, and Weiner

2014). It will be important to know whether immune responses to proto-tumors trade off with other functions. Alternatively, we may rely primarily on cancer-preventing maintenance mechanisms operated by multiple tumor-suppressor and DNA repair pathways. The function of some of these mechanisms may decline with age. That would explain the age-associated increase in cancer incidence and raises two questions: do these mechanisms trade off with other functions, and why do they decline with age?

Is there a connection between invasive placentas and metastasis? Many cancers originate in stem cells, and some stem cells have adaptations that especially predispose to metastasis. The embryonic stem cells in species with invasive placentas invade the endometrium and insert themselves into maternal tissue, including into the walls of maternal arteries: they are capable of moving into and colonizing foreign tissue. That capacity lies dormant in differentiated tissue but can be reawakened by an appropriate set of mutations that recruit it into metastatic performance (Murray and Lessey 1999; Brüning, Makovitzky, and Gingelmaier 2009). We now know that some animals with less invasive placentation have less risk of metastatic cancer than animals with more invasive placentation, including ourselves (see Figure 7.12). We need more such comparisons and further studies that compare gene expression in cells invading endometria with gene expression in metastasizing cells. An early example of such a study showed some promise (Ma et al. 2011).

How does natural selection shape tumor growth and metastasis? What are the traits that allow cancer cells to succeed, and which genes are the ones whose expression shapes those traits? Recent research provides a framework in which to pursue those key issues (P. J. Campbell et al. 2010; Gerlinger and Swanton 2010; Tian et al. 2011; Ding et al. 2012; Gerlinger et al. 2012); understanding how selection produces superior growth and metastasis could identify biomarkers that would aid in reliable early detection of cancer.

Do early life history–late life history trade-offs suggest treatments for aging?

In mammals, the target of rapamycin cellular signaling network mediates trade-offs between performance early and late in life, possibly including the risk of Alzheimer's disease (Caccamo et al. 2014). This finding suggested that rapamycin supplements might extend life, and they do so in mice, but without the effect on health that would also be desired (Johnson et al. 2013; Neff et al. 2013). The idea that prompted this line of research started with a question: what mechanisms mediate the trade-offs between early and late life? Other such mechanisms doubtless exist, and finding them could suggest other, more successful treatments that extend life.

What is the role of parent-of-origin imprinting in mental disease?

It is now well established that parent-of-origin imprinting helps mediate conflicts between mother and father over maternal investment in the fetus

(Haig 1993, 2000, 2004, 2010). Another set of genes with either parent-of-origin imprinting or variation in copy number are expressed in the brain, where they appear to be involved in a conflict over postpartum offspring behavior (Haig and Wharton 2003). Usually, that conflict yields an intermediate result and a healthy child, but some researchers (Crespi and Badcock 2008; Crespi, Stead, and Elliot 2010) suggest that when it is disrupted by mutation or environmental insult, mental disease can arise. They see the disruption of the normal state as revealing an axis along which autism and schizophrenia form the extremes. When the father's interests are overexpressed and the mother's are underexpressed, incremental effects increase the risk of autism in extreme cases. When the mother's interests are overexpressed and the father's are underexpressed, incremental effects increase the risk of schizophrenia in extreme cases. Less extreme disruptions produce intermediate impact on mental function, including autism-spectrum disorders and mild psychoses. Support for this idea comes from a study that analyzed the risk of mental disorders in 1.75 million children born in Denmark between 1977 and 2009 (see Chapter 9).

Both autism and schizophrenia have many causes, and if this idea is shown to work, it will become part of a multicausal explanation. If confirmed, it will remarkably connect evolutionary conflict theory to mental disease, an insight into serious medical disorders from a completely unexpected direction. Independent, skeptical groups should test the hypothesis, and if the results confirm the predictions, the mechanisms that produce them should be identified.

■ SUMMARY

Evolutionary medicine is a relatively new field in which some issues are well supported by reliable results and others remain open. Some of the open issues are pressing and deserve immediate attention and investment.

Why some issues were not addressed in more detail

Much of the science that supports evolutionary medicine is strong, and that is the part we have emphasized in this book. However, some of it is weak, and some of it is biased. Here we call attention to several places where claims are not supported by evidence.

Life history evolution does not predict intelligence or criminality

Canadian psychologist Philip Rushton (2000) claims that life history theory can be used to predict differences in the intelligence and criminal tendencies of races. In particular, he asserts that, in contrast to Asians and whites, blacks share an evolved, inherited fast life history with lower intelligence and greater risk of criminality. His work is a prime example of confusing correlation with causation because the data on which he bases his claims are seriously confounded by poverty and discrimination. In particular, there is no study in which other ethnicities have been exposed to the same levels of poverty and discrimination as have blacks to see what the outcomes would be for intelligence and criminality. Everything that we know about the genetics of ethnicity (see Chapter 2) suggests that the correlations Rushton reports are readily explained by differences in environment, not genes. For that reason, we have avoided further comment on differences among ethnicities in life history and behavior. Ethnicities do differ in risk of adverse drug responses, but that has nothing to do with intelligence or behavior.

Much of what is written about paleodiets is bad science

The paleodiet movement is probably having a constructive influence on health and longevity, but much of the science on which some of its claims are based is flawed. Its arguments invoke mismatches between modern diets and the diets our Paleolithic ancestors experienced prior to the agricultural revolution about 10,000 years ago. Although it is certainly true that we now have much more refined grain, fatty meat, and sugar and less raw fruits, vegetables, and tubers in our diets than did our ancestors, it is not clear that the change in diet is the principal cause of obesity, cancer, or cardiovascular disease. Among the reasons to be skeptical of that claim, these two seem to us to be most important:

1. Our ancestors inhabited many environments and experienced many diets, some of them with no fresh fruits or vegetables and with lots of meat and fat (e.g., the Inuit diet). There was no single Paleolithic diet that could serve as a reference point; our coevolution with our diets, while lagging them, has been continuous; and some of the diets our ancestors ate would be characterized by paleodiet advocates as decidedly unhealthy.
2. Many of the conditions attributed to the change in diet are caused in significant part simply because we now live longer. We do not have good data for the appropriate comparison: risks of disease and death in modern populations of people eating paleodiets versus those eating diets high in refined grains, sugar, and fatty meats who have all been released from the threats of infectious disease, predation, and death in childbirth.

We do not deny that our appetite for sweets and carbohydrates may have been shaped by an evolutionary past in which they were scarce, nor do we deny the possibility that the current epidemic of obesity and diabetes may be mediated in part by a mismatch between our evolved preferences and our new obesogenic environment. We simply point out that the science supporting such claims is not yet solid and that we do not need to invoke evolution to eat a more healthful diet and to exercise more.

■ SUMMARY

Evolutionary thinking is seductive because it provides powerful explanations, but it is easily misapplied, especially by those not trained in the skepticism that is part of standard scientific research. There are many claims about diet and psychology that do not meet the best standards of experimental and clinical science. Their status should not be allowed to call into question the parts of evolutionary medicine on which we have concentrated, those where the science is much more solid.

Differences between classical and evolutionary medicine

First, evolutionary medicine does not replace classical medicine; rather, it complements it by giving additional insights. Those insights are much more important in some places, such as infectious disease, cancer, and autoimmune disease, than they are in others.

Probably the most important general difference concerns the basic assumptions that implicitly drive decisions. There is a tendency in modern molecular medicine to regard patients as machines with parts that can be replaced if broken and pathogens as machines that can be broken to disable them, without other consequences. Evolutionary medicine sees both patients and pathogens as bundles of trade-offs in which it is impossible to change one thing without also changing something else, and because evolution has already greatly improved fitness in the face of trade-offs and constraints, making it likely that any change will reveal costs, these changes are often for the worse. This insight does not mean that we should not treat, but it does mean that we should have heightened awareness about the consequences of treatment.

We should remain very aware that pathogens and cancer cells have their own evolutionary agendas and that they respond rapidly to any human intervention with countermeasures. We need to think creatively about antimicrobial and anticancer therapies that are evolution-proof; if we do not solve that problem, we will be faced with a world in which infection is much less easily treated than it could be, operations are therefore much riskier than they should be, and patients die sooner than necessary from cancer. Classical medicine has been aware of these issues, but only evolutionary medicine has opened the possibility of treatments that do not produce an evolved response or, if they do, slow it down significantly.

It had certainly not been standard in obstetrics to think of the fetus and infant as being in conflict with the mother about maternal investment or to see that some of infant behavior could be mediated by conflicts between maternal and paternal interests. Evolutionary conflict theory has remodeled our views of family life in fundamental ways, some of which are starting to have payoffs in terms of fresh insights into medical conditions.

The role of mismatch in generating diseases of civilization is probably best established wherever disturbances to the microbiome mediate abnormal responses by the immune system. There the guiding evolutionary principle is not current natural selection shaping adaptations; rather, it is past evolution delivering inherited constraints that do not yet function optimally in a rapidly changing environment. There are many other ways we are mismatched to modern environments in which diet, exercise, light, landscapes, group size, social units, and other features have been drastically altered. It will take a great deal of research to work out the implications of these mismatches for health and disease.

■ SUMMARY

Evolutionary approaches to medicine complement classical views; they do not replace them. It would be foolish to displace fascinating and useful molecular and cell science with knowledge derived from evolutionary biology. Instead, we should connect the insights of evolutionary biology to those of molecular and cell science to produce new interdisciplinary research and new integrated knowledge that is exciting, general, and useful. It will reduce suffering and save lives.

References

Ackermann, M., R. Bijlsma, A. C. James, et al. 2001. Effects of assay conditions in life history experiments with *Drosophila melanogaster*. *Journal of Evolutionary Biology* 14: 199–209.

Ackermann, M., S. C. Stearns, and U. Jenal. 2003. Senescence in a bacterium with asymmetric division. *Science* 300: 1920–1920.

Agundez, J. A. G. 2008. Polymorphisms of human n-acetyltransferases and cancer risk. *Current Drug Metabolism* 9: 520–531.

Alekshun, M. N., and S. B. Levy. 2007. Molecular mechanisms of antibacterial multidrug resistance. *Cell* 128: 1037–1050.

Allen, R. C. 2014. Poverty lines in history, theory, and current international practice. Discussion Paper Series 685, Department of Economics, Oxford University.

Allison, A. C. 1964. Polymorphism and natural selection in human populations. *Cold Spring Harbor Symposia on Quantitative Biology* 29: 137–149.

American Cancer Society. 2006. *Cancer Facts and Figures 2006*. American Cancer Society, Atlanta.

American Society for Reproductive Medicine, Practice Committee. 2008. Hormonal contraception: recent advances and controversies. *Fertility and Sterility* 90: S103–S113.

Anderson, K., C. Lutz, F. W. van Delft, et al. 2011. Genetic variegation of clonal architecture and propagating cells in leukaemia. *Nature* 469: 356–362.

Anderson, R. M., and R. M. May. 1985. Vaccination and herd immunity to infectious diseases. *Nature* 318: 323–329.

Archetti, M. 2013. Evolutionarily stable anticancer therapies by autologous cell defection. *Evolution, Medicine, and Public Health* 2013: 161–172.

Archetti, M., D. A. Ferraro, and G. Christofori, 2015. Heterogeneity for IGF-II production maintained by public goods dynamics in neuroendocrine pancreatic cancer. *Proceedings of the National Academy of Sciences, USA* 112: 1833–1838.

Arenz, S., R. Ruckerl, B. Koletzko, et al, 2004. Breast-feeding and childhood obesity—A systematic review. *International Journal of Obesity* 28: 1247–1256.

Armstrong, G. L., L. A. Conn, and R. W. Pinner. 1999. Trends in infectious disease mortality in the United States during the 20th century. *JAMA : The Journal of the American Medical Association* 281: 61–66.

Austad, S., and C. E. Finch. 2008. The evolutionary context of human aging and degenerative disease. In *Evolution in Health and Disease*, S. C. Stearns and J. C. Koella (eds.), Oxford University Press, Oxford and New York, pp. 301–312.

Bach, J. 2002. The effect of infections on susceptibility to autoimmune and allergic diseases. *New England Journal of Medicine* 347: 911–920.

Bager, P., J. Wohlfahrt, and T. Westergaard. 2008. Caesarean delivery and risk of atopy and allergic disesase: Meta-analyses. *Clinical and Experimental Allergy* 38: 634–642.

Balbo, N., F. Billari, and M. Mills. 2013. Fertility in advanced societies: A review of research. *European Journal of Population* 29: 1–38.

Barbujani, G., and L. Excoffier. 1999. The history and geography of human genetic diversity. In *Evolution in Health and Disease*, S. C. Stearns and J. C. Koella (eds.), Oxford University Press, Oxford, pp. 27–39.

Barbujani, G., A. Magagni, E. Minch, et al. 1997. An apportionment of human DNA diversity. *Proceedings of the National Academy of Sciences, USA* 94: 4516–4519.

Barker, D. J. P., C. Osmond, T. J. Forsén, et al. 2005. Trajectories of growth among children who have coronary events as adults. *The New England Journal of Medicine* 353: 1802–1809.

Barker, D. J. P., P. D. Winter, C. Osmond, et al. 1989. Weight in infancy and death from ischemic heart-disease. *The Lancet* 2: 577–580.

Beall, C. M. 2006. Andean, Tibetan, and Ethiopian patterns of adaptation to high-altitude hypoxia. *Integrative and Comparative Biology* 46: 18–14.

Beall, C. M., G. L. Cavalleri, L. Deng, et al. 2010. Natural selection on *EPAS1 (HIF2a)* associated with low hemoglobin concentration in Tibetan highlanders. *Proceedings of the National Academy of Sciences USA* 107: 11459–11464.

Belkaid, Y., and J. A. Segre. 2014. Dialogue between skin microbiota and immunity. *Science* 346: 954–959.

Berenos, C., P. Schmid-Hempel, and K. Wegner. 2009. Evolution of host resistance and trade-offs between virulence and transmission potential in an obligately killing parasite. *Journal of Evolutionary Biology* 22: 2049–2056.

Bergstrom, C., and M. Feldgarden. 2008. The ecology and evolution of antibiotic-resistant bacteria. In *Evolution in Health and Disease*, S. C. Stearns and J. C. Koella, (eds.), Oxford University Press, Oxford and New York, pp. 125–137.

Bongaarts, J. 2001. Fertility and reproductive preferences in post-transitional societies. *Population and Development Review* 27: 260.

Bowen, R. 2011. "Placental structure and classification." Colorado State University. http://www.vivo.colostate.edu/hbooks/pathphys/reprod/placenta/structure.html.

Bråbäck, L. L., H. H. Vogt, and A. A. Hjern. 2011. Migration and asthma medication in international adoptees and immigrant families in Sweden. *Clinical and Experimental Allergy* 41: 1108–1115.

Brown, G. R., K. N. Laland, and M. B. Mulder. 2009. Bateman's principles and human sex roles. *Trends in Ecology & Evolution* 24: 297–304.

Brown, S. P., S. A. West, S. P. Diggle, et al. 2009. Social evolution in micro-organisms and a Trojan horse approach to medical intervention strategies. *Philosophical Transactions of the Royal Society B, Biological Sciences* 364: 3157–3168.

Brundage, J. F., and G. D. Shanks. 2008. Deaths from bacterial pneumonia during 1918–19 influenza pandemic. *Emerging Infectious Diseases* 14: 1193–1199.

Brüning, A., J. Makovitzky, and A. Gingelmaier. 2009. The metastasis-associated genes MTA1 and MTA3 are abundantly expressed in human placenta and chorionic carcinoma cells. *Histochemistry and Cell Biology* 132: 33–38.

Bull, J. J., and A. S. Lauring. 2014. Theory and empiricism in virulence evolution. *PLoS Pathogens* 10: e1004387.

Byars, S. G., D. Ewbank, D. R. Govindaraju, et al. 2010. Natural selection in a contemporary human population. *Proceedings of the National Academy of Sciences, USA* 107: 1787–1792.

Byars, S. G., S. C. Stearns, and J. J. Boomsma. 2014. Opposite risk patterns for autism and schizophrenia are associated with normal variation in birth size: Phenotypic support for hypothesized diametric gene-dosage effects. *Proceedings of the Royal Society of London, Series B, Biological Sciences* 281: 20140604.

Caccamo, A., V. De Pinto, A. Messina, et al. 2014. Genetic reduction of mammalian target of rapamycin ameliorates Alzheimer's disease-like cognitive and pathological deficits by restoring hippocampal gene expression signature. *Journal of Neuroscience* 34: 7988–7998.

Caldwell, J. 1999. Paths to lower fertility. *BMJ: British Medical Journal* 319: 985–987.

Campbell, K., and J. Wood. 1988. Fertility in transitional societies. In *Natural Human Fertility: Social and Biological Determinants*, P. Diggory, M. Potts, and S. Teper (eds.), Macmillan, London, pp. 33–69.

Campbell, P. J., S. Yachida, L. J. Mudie, et al. 2010. The patterns and dynamics of genomic instability in metastatic pancreatic cancer. *Nature* 467: 1109–1113.

Capellini, I., C. Venditti, and R. A. Barton. 2011. Placentation and maternal investment in mammals. *The American Naturalist* 177: 86–98.

Carroll, S. B., J. K. Grenier, and S. D. Weatherbee. 2005. *From DNA to Diversity: Molecular Genetics and the Evolution of Animal Design,* 2nd ed. Blackwell Science, Malden, MA.

Carter, A. M., and R. Pijnenborg. 2011. Evolution of invasive placentation with special reference to non-human primates. *Best Practice & Research: Clinical Obstetrics & Gynaecology* 25: 249–257.

Cascorbi, I., I. Roots, and J. Brockmöller. 2001. Association of NAT1 and NAT2 polymorphisms to urinary bladder cancer: Significantly reduced risk in subjects with NAT1*10. *Cancer Research* 61: 5051–5056.

Cerf-Bensussan, N., and V. Gaboriau-Routhiau. 2010. The immune system and the gut microbiota: Friends or foes? *Nature Reviews Immunology* 10: 735–744.

Chahroudi, A., S. E. Bosinger, T. H. Vanderford, et al. 2012. Natural SIV hosts: Showing AIDS the door. *Science* 335: 1188–1193.

Chapman, S. J., and A. Hill. 2012. Human genetic susceptibility to infectious disease. *Nature Reviews Genetics* 13: 175–188.

Chibeu, A. A., E. J. E. Lingohr, L. L. Masson, et al. 2012. Bacteriophages with the ability to degrade uropathogenic *Escherichia coli* biofilms. *Viruses* 4: 471–487.

Chichlowski, M., G. De Lartigue, J. B. German, et al. 2012. Bifidobacteria isolated from infants and cultured on human milk oligosaccharides affect intestinal epithelial function. *Journal of Pediatric Gastroenterology and Nutrition* 55: 321–327.

Cho, J. H. 2008. The genetics and immunopathogenesis of inflammatory bowel disease. *Nature Reviews Immunology* 8: 458–466.

Clarke, G. 2008. *A Farewell to Alms: A Brief Economic History of the World.* Princeton University Press, Princeton.

Coale, A. 1986. The decline of fertility in Europe since the eighteenth century as a chapter in demographic history. In *The Decline of Fertility in Europe,* A. Coale and S. Watkins (eds.), Princeton University Press, Princeton, NJ, pp. 1–3 .

Cooper, M. D., and M. N. Alder. 2006. The evolution of adaptive immune systems. *Cell* 124: 815–822.

Corbett, S., A. Courtiol, V. Lummaa, et al. 2015, in press. A fiery forge—The industrial revolution, recent human evolution, and the global burden of non-communicable disease. *Proceedings of the National Academy of Sciences, USA.*

Correale, J., and M. F. Farez. 2011. The impact of parasite infections on the course of multiple sclerosis. *Journal of Neuroimmunology* 233: 6–11.

Courtiol, A., I. J. Rickard, V. Lummaa, et al. 2013. The demographic transition influences variance in fitness and selection on height and BMI in rural Gambia. *Current Biology* 23: 884–889.

Crespi, B., and C. Badcock. 2008. Psychosis and autism as diametrical disorders of the social brain. *Behavioral and Brain Sciences* 31: 241–320.

Crespi, B., and K. Summers. 2005. Evolutionary biology of cancer. *Trends in Ecology & Evolution* 20: 545–552.

Crespi, B., P. Stead, and M. Elliot. 2010. Comparative genomics of autism and schizophrenia. *Proceedings of the National Academy of Sciences, USA* 107(Suppl. 1): 1736–1741.

Crosley, E. J., M. G. Elliot, J. K. Christians, et al. 2013. Placental invasion, preeclampsia risk and adaptive molecular evolution at the origin of the great apes: Evidence from genome-wide analyses. *Placenta* 34: 127–132.

Crow, J. 1958. Some possibilities for measuring selection intensities in man. *Human Biology* 30: 1–13.

Daan, S., C. Dijkstra, and J. M. Tinbergen. 1990. Family planning in the kestrel (*Falco tinnunculus*): The ultimate control of covariation of laying date and clutch size. *Behaviour* 114: 83–116.

Daly, A. K. 2010. Genome-wide association studies in pharmacogenomics. *Nature Reviews Genetics* 11: 241–246.

Darwin, C. 2002. Journal and remarks, 1832–1836. In *Charles Darwin in Australia,* F. Nicholas and J. Nicholas (eds.), Cambridge University Press, pp. 30–31.

de Jong, G., and A. J. van Noordwijk. 1992. Acquisition and allocation of resources: Genetic (CO) variances, selection, and life histories. *The American Naturalist* 139: 749–770.

Ding, L., T. J. Ley, D. E. Larson, et al. 2012. Clonal evolution in relapsed acute myeloid leukaemia revealed by whole-genome sequencing. *Nature* 481: 506–510.

Di Rienzo, A. 2006. Population genetics models of common diseases. *Current Opinion in Genetics and Development* 16: 630–636.

D'Souza, A. W., and G. P. Wagner. 2014. Malignant cancer and invasive placentation: A case for positive pleiotropy between endometrial and malignancy phenotypes. *Evolution, Medicine, and Public Health* 2014: 136–145.

Dyson, T., and M. Murphy. 1985. The onset of fertility transition. *Population and Development Review* 11: 399.

Ebert, D. 1998. Evolution—Experimental evolution of parasites. *Science* 282: 1432–1435.

Elliott, D. E. D., R. W. R. Summers, and J. V. J. Weinstock. 2007. Helminths as governors of immune-mediated inflammation. *International Journal for Parasitology* 37: 8.

Ellison, P. T. 2003. Energetics and reproductive effort. *American Journal of Human Biology* 15: 342–351.

Ellison, P. T. 2009. *On Fertile Ground*. Harvard University Press, Cambridge, MA.

Ellison, P. T., and M. A. Ottinger. 2014. A comparative perspective on reproductive aging, reproductive cessation, post-reproductive life, and social behavior. In *Sociality, Hierarchy, Health: Comparative Biodemography: Papers from a Workshop*, M. Weinstein and M. A. Lane (eds.), National Academy of Science, Washington, DC, pp. 315–338.

Emera, D., R. Romero, and G. Wagner. 2011. The evolution of menstruation: A new model for genetic assimilation. *BioEssays* 34: 26–35.

Enattah, N. S., A. Trudeau, V. Pimenoff, et al. 2007. Evidence of still-ongoing convergence evolution of the lactase persistence T-13910 alleles in humans. *American Journal of Human Genetics* 81: 615–625.

Enoch, M.-A. 2014. Genetic influences on response to alcohol and response to pharmacotherapies for alcoholism. *Pharmacology, Biochemistry, and Behavior* 123: 17–24.

Farooqi, I. S., and J. M. Hopkin. 1998. Early childhood infection and atopic disease. *Thorax* 53: 927–932.

Fenner, F. 1983. The Florey lecture, 1983. Biological control, as exemplified by smallpox eradication and myxomatosis. *Proceedings of the Royal Society of London, Series B, Biological Science* 218: 259–285.

Ferretti, C., L. Bruni, V. Dangles-Marie, et al. 2007. Molecular circuits shared by placental and cancer cells, and their implications in the proliferative, invasive and migratory capacities of trophoblasts. *Human Reproduction Update* 13: 121–141.

Finch, C. E., and E. M. Crimmins. 2004. Inflammatory exposure and historical changes in human life-spans. *Science* 305: 1736–1739.

Floud, R., R. Fogel, B. Harris, et al. 2011. *The Changing Body: Health, Nutrition and Human Development in the Western Worlds since 1700*. Cambridge University Press, Cambridge.

Fogel, R. 2004. *The Escape from Hunger and Premature Mortality 1750–2100*. Cambridge University Press, Cambridge.

Frisch, R. E. 1978. Population, food intake, and fertility. There is historical evidence for a direct effect of nutrition on reproductive ability. *Science* 199: 22–30.

Froehlich, J. W. 1970. Migration and the plasticity of physique in the Japanese-Americans of Hawaii. *American Journal of Physical Anthropology* 32: 429–442.

Gandon, S., M. J. Mackinnon, S. Nee, et al. 2001. Imperfect vaccines and the evolution of pathogen virulence. *Nature* 414: 751–756.

Gatenby, R. A., A. S. Silva, R. J. Gillies, et al. 2009. Adaptive therapy. *Cancer Research* 69: 4894–4903.

Gause, W. C., T. A. Wynn, and J. E. Allen. 2013. Type 2 immunity and wound healing: Evolutionary refinement of adaptive immunity by helminths. *Nature Reviews Immunology* 13: 607–614.

Gebhardt, M. D., and S. C. Stearns. 1988. Reaction norms for developmental time and weight at eclosion in *Drosophila mercatorum*. *Journal of Evolutionary Biology* 1: 335–354.

Gerlinger, M., and C. Swanton. 2010. How Darwinian models inform therapeutic failure initiated by clonal heterogeneity in cancer medicine. *British Journal of Cancer* 103: 1139–1143.

Gerlinger, M., A. J. Rowan, S. Horswell, et al. 2012. Intratumor heterogeneity and branched evolution revealed by multiregion sequencing. *The New England Journal of Medicine* 366: 883–892.

Gilbert, S. F. 2000. *Developmental Biology*, 6th edition. Sinauer, Sunderland, MA.

Gilbert, S. F., and D. Epel. 2015. *Ecological Developmental Biology: The Environmental Regulation of Development, Health, and Evolution*. 2nd edition. Sinauer, Sunderland, MA.

Gillies, R. J., D. Verduzco, and R. A. Gatenby. 2012. Evolutionary dynamics of carcinogenesis and why targeted therapy does not work. *Nature Reviews Cancer* 12: 487–493.

Glocker, E.-O., D. Kotlarz, K. Boztug, et al. 2009. Inflammatory bowel disease and mutations affecting the interleukin-10 receptor. *New England Journal of Medicine* 361: 2033–2045.

Godfrey, K. M., A. Sheppard, P. D. Gluckman, et al. 2011. Epigenetic gene promoter methylation at birth is associated with child's later adiposity. *Diabetes* 60: 1528–1534.

Gögele, M., C. Pattaro, C. Fuchsberger, et al. 2011. Heritability analysis of life span in a semi-isolated population followed across four centuries reveals the presence of pleiotropy between life span and reproduction. *The Journals of Gerontology Series A: Biological Sciences and Medical Sciences* 66: 26–37.

Gojobori, T., E. N. Moriyama, and M. Kimura. 1990. Molecular clock of viral evolution, and the neutral theory. *Proceedings of the National Academy of Sciences, USA* 87: 10015–10018.

Goldman, D., G. Oroszi, and F. Ducci. 2005. The genetics of addictions: Uncovering the genes. *Nature Reviews Genetics* 6: 521–532.

Greaves, M., and C. C. Maley. 2012. Clonal evolution in cancer. *Nature* 481: 306–313.

Guégan, J.-F., F. Prugnolle, and F. Thomas. 2009. Global spatial patterns of infectious disease and human evolution. In *Evolution in Health and Disease*, S. Stearns and J. Koella (eds.), Oxford University Press, Oxford, pp. 19–29.

Guégan, J.-F., F. Thomas, M. E. Hochberg, et al. 2001. Disease diversity and human fertility. *Evolution* 55: 1308–1314.

Hagen, E. H., R. J. Sullivan, R. Schmidt, et al. 2009. Ecology and neurobiology of toxin avoidance and the paradox of drug reward. *Neuroscience* 160: 69–84.

Haig, D. 1993. Genetic conflicts in human pregnancy. *Quarterly Review of Biology* 68: 495–532.

Haig, D. 2000. The kinship theory of genomic imprinting. *Annual Review of Ecology and Systematics* 31: 9–32.

Haig, D. 2004. Genomic imprinting and kinship: How good is the evidence? *Annual Review of Genetics* 38: 553–585.

Haig, D. 2010. Transfers and transitions: Parent-offspring conflict, genomic imprinting, and the evolution of human life history. *Proceedings of the National Academy of Sciences, USA* 107: 1731–1735.

Haig, D. 2014. Troubled sleep: Night waking, breastfeeding, and parent–offspring conflict. *Evolution, Medicine, and Public Health* 2014(1): 32–39.

Haig, D., and R. Wharton. 2003. Prader-Willi syndrome and the evolution of human childhood. *American Journal of Human Biology* 15: 320–329.

Hales, C. N., and D. J. Barker. 1992. Type 2 (non-insulin-dependent) diabetes mellitus: The thrifty phenotype hypothesis. *Diabetologia* 35: 595–601.

Hamilton, W. D. 1964a. The genetical evolution of social behaviour. I. *Journal of Theoretical Biology* 7: 1–16.

Hamilton, W. D. 1964b. The genetical evolution of social behaviour. II. *Journal of Theoretical Biology* 7: 17–52.

Han, Y., S. Gu, H. Oota, et al. 2007. Evidence of positive selection on a class I ADH locus. *American Journal of Human Genetics* 80: 441–456.

Harper, K. N., M. K. Zuckerman, M. L. Harper, et al. 2011. The origin and antiquity of syphilis revisited: An appraisal of Old World pre-Columbian evidence for treponemal infection. *American Journal of Physical Anthropology* 54: 99–133.

Harvey, P. H., and T. H. Clutton-Brock. 1985. Life history variation in primates. *Evolution* 39: 559.

Hawkes, K., J. F. O'Connell, N. Jones, et al. 1998. Grandmothering, menopause, and the evolution of human life histories. *Proceedings of the National Academy of Sciences, USA* 95: 1336–1339.

Heyer, E., L. Brazier, L. Ségurel, et al. 2011. Lactase persistence in central Asia: Phenotype, genotype, and evolution. *Human Biology* 83: 379–392.

Hilborn, R., and S. C. Stearns. 1982. On inference in ecology and evolutionary biology—The problem of multiple causes. *Acta Biotheoretica* 31: 145–164.

Hill, A. V. S. 2012. Evolution, revolution and heresy in the genetics of infectious disease susceptibility. *Philosophical Transactions of the Royal Society B, Biological Sciences* 367: 840–849.

Holliday, R. 1989. Food reproduction and longevity: Is the extended lifespan of calorie restricted animals an evolutionary adaptation? *Bioessays* 10: 125–127.

Huh, S. Y. S., S. L. S. Rifas-Shiman, C. A. C. Zera, et al. 2012. Delivery by caesarean section and risk of obesity in preschool age children: A prospective cohort study. *Archives of Disease in Childhood* 97: 610–616.

Huijben, S., A. S. Bell, D. G. Sim, et al. 2013. Aggressive chemotherapy and the selection of drug resistant pathogens. *PLoS Pathogens* 9: e1003578.

Izcue, A., J. L. Coombes, and F. Powrie. 2009. Regulatory lymphocytes and intestinal inflammation. *Annual Review of Immunology* 27: 313–338.

Jablonski, N. G., and G. Chaplin. 2010. Colloquium paper: Human skin pigmentation as an adaptation to UV radiation. *Proceedings of the National Academy of Sciences, USA* 107(Suppl. 2): 8962–8968.

Jacquelin, B., V. Mayau, B. Targat, et al. 2009. Nonpathogenic SIV infection of African green monkeys induces a strong but rapidly controlled type I IFN response. *Journal of Clinical Investigation* 119: 3544–3555.

Janssen, R., B. J. Eriksson, N. N. Tait, et al. 2014. Onychophoran Hox genes and the evolution of arthropod Hox gene expression. *Frontiers in Zoology* 11: 22.

Johnson, S. C., G. M. Martin, P. S. Rabinovitch, et al. 2013. Preserving youth: Does rapamycin deliver? *Science Translational Medicine* 5: 211fs240.

Kang, H.-J., H. Feng, Y. Sun, et al. 2009. Single-nucleotide polymorphisms in the p53 pathway regulate fertility in humans. *Proceedings of the National Academy of Sciences, USA* 106: 9761–9766.

Kareva, I., and P. Hahnfeldt. 2013. The emerging "hallmarks" of metabolic reprogramming and immune evasion: Distinct or linked? *Cancer Research* 73: 2737–2742.

Kimura, M. 1983. *The Neutral Theory of Molecular Evolution*. Cambridge University Press, Cambridge.

Kirkwood, T. B., and T. Cremer. 1982. Cyto-gerontology since 1881: A reappraisal of August Weismann and a review of modern Progress. *Human Genetics* 60: 101–121.

Kirkwood, T. B., and R. Holliday. 1979. The evolution of ageing and longevity. *Proceedings of the Royal Society of London, Series B, Biological Sciences* 205: 531–546.

Kirkwood, T. B., and D. P. Shanley. 2005. Food restriction, evolution and ageing. *Mechanisms of Ageing and Development* 126: 1011–1016.

Koella, J., and L. Lorenz. 2009. Microsporidians as evolution-proof agents of malaria control? *Advances in Parasitology* 68: 315–327.

Koella, J., P. Lynch, and M. Thomas. 2009. Towards evolution-proof malaria control with insecticides. *Evolutionary Applications* 2: 469–480.

Kohler, H., J. Rodgers, and K. Christensen. 2002. Between nurture and nature: The shifting determinants of female fertility in Danish twin cohorts. *Social Biology* 3/4: 218–248.

Kohn, G. C. 2008. *Encyclopedia of Plague and Pestilence: From Ancient Times to the Present*. 3rd edition. Choice Reviews Online, New York.

Kondrashova, A., T. Seiskari, J. Ilonen, et al. 2012. The 'Hygiene hypothesis' and the sharp gradient in the incidence of autoimmune and allergic diseases between Russian, Karelia, and Finland. *APMIS : Acta Pathologica, Microbiologica, et Immunologica Scandinavica* 121: 478–493.

Kono, T. 2001. *Weightlifting, Olympic Style*. Hawaii Kono Company, Oahu, HI.

Kostadinov, R. L., M. K. Kuhner, X. Li, et al. 2013. NSAIDs modulate clonal evolution in Barrett's Esophagus. *PLoS Genetics* 9: e1003553.

Kuningas, M., S. Altmäe, A. G. Uitterlinden, et al. 2011. The relationship between fertility and lifespan in humans. *Age* 33: 615–622

Lahdenperä, M., D. O. S. Gillespie, V. Lummaa, et al. 2012. Severe intergenerational reproductive conflict and the evolution of menopause. *Ecology Letters* 15: 1283–1290.

Lahdenperä, M., V. Lummaa, S. Helle, et al. 2004. Fitness benefits of prolonged postreproductive lifespan in women. *Nature* 428: 178–181.

Lahdenperä, M., A. F. Russell, M. Tremblay, et al. 2011. Selection on menopause in two premodern human populations: No evidence for the Mother Hypothesis. *Evolution* 65: 476–489.

Lee, R. D. 2003a. The demographic transition: Three centuries of fundamental change. *Journal of Economic Perspectives* 17: 167–190.

Lee, R. D. 2003b. Rethinking the evolutionary theory of aging: Transfers, not births, shape senescence in social species. *Proceedings of the National Academy of Sciences, USA* 100: 9637–9642.

Levin, B. R., and J. J. Bull. 2004. Population and evolutionary dynamics of phage therapy. *Nature Reviews Microbiology* 2: 166–173.

Levin, B. R., V. Perrot, and N. Walker. 2000. Compensatory mutations, antibiotic resistance and the population genetics of adaptive evolution in bacteria. *Genetics* 154: 985–997.

Lewontin, R. 1974. *The Genetic Basis of Evolutionary Change.* Columbia University Press, New York.

Lewontin, R. 1995. The apportionment of human diversity. In *Evolutionary Biology*, T. Dobzhansky, M. K. Hecht, and W. C. Steere (eds.), Springer, Boston, pp. 381–398.

Li, J. Z., D. M. Absher, H. Tang, et al. 2008. Worldwide human relationships inferred from genome-wide patterns of variation. *Science* 319: 1100–1104.

Lieberman, D. E. 2013. *The Story of the Human Body: Evolution, Health, and Disease.* Vintage Books, New York.

Lion, E., Y. Willemen, Z. N. Berneman, et al. 2012. Natural killer cell immune escape in acute myeloid leukemia. *Leukemia* 26: 2019–2026.

Liu, V. C., L. Y. Wong, T. Jang, et al. 2007. Tumor evasion of the immune system by converting CD4+ CD25− T cells into CD4+ CD25+ T regulatory cells: Role of tumor-derived TGF-β. *The Journal of Immunology* 178: 2883–2892.

Loc-Carrillo, C., and S. T. Abedon. 2011. Pros and cons of phage therapy. *Bacteriophage* 1: 111–114.

Loisel, D. A., S. C. Alberts, and C. Ober. 2008. Functional significance of MHC variation in mate choice, reproductive outcome, and disease risk. In *Evolution in Health and Disease*, S. C. Stearns and J. C. Koella (eds.), Oxford University Press, Oxford and New York, pp. 95–108.

Loudon, I. 2000. Maternal mortality in the past and its relevance to developing countries today. *The American Journal of Clinical Nutrition* 72: 241S–246S.

Lozano, R., M. Naghavi, K. Foreman, et al. 2012. Global and regional mortality from 235 causes of death for 20 age groups in 1990 and 2010: A systematic analysis for the Global Burden of Disease Study 2010. *The Lancet* 380: 2095–2128.

Lozupone, C. A., J. I. Stombaugh, J. I. Gordon, et al. 2012. Diversity, stability and resilience of the human gut microbiota. *Nature* 489: 220–230.

Lu, L., D. F. Mackay, and J. P. Pell. 2013. Association between level of exposure to secondhand smoke and peripheral arterial disease: Cross-sectional study of 5686 never smokers. *Atherosclerosis* 229: 273–276.

Luksza, M., and M. Lässig. 2014. A predictive fitness model for influenza. *Nature* 507: 57–61.

Lummaa, V., J. Jokela, and E. Haukioja. 2001. Gender difference in benefits of twinning in pre-industrial humans: Boys did not pay. *Journal of Animal Ecology* 70: 739–746.

Lynch, V. J., and G. P. Wagner. 2005. The birth of the uterus. *Natural History* 114: 36–41.

Lynch, V. J., R. D. Leclerc, G. May, et al. 2011. Transposon-mediated rewiring of gene regulatory networks contributed to the evolution of pregnancy in mammals. *Nature Genetics* 43: 1154–1159.

Ma, Y.-L., P. Zhang, F. Wang, et al. 2011. Human embryonic stem cells and metastatic colorectal cancer cells shared the common endogenous human microRNA-26b. *Journal of Cellular and Molecular Medicine* 15: 1941–1954.

Mackinnon, M. J., and A. F. Read. 2004. Virulence in malaria: An evolutionary viewpoint. *Philosophical Transactions of the Royal Society B, Biological Sciences* 359: 965–986.

Mackinnon, M. J., S. Gandon, and A. F. Read. 2008. Virulence evolution in response to vaccination: The case of malaria. *Vaccine* 26(Suppl. 3): C42–C52.

Madrigal, L., and M. Meléndez-Obando. 2008. Grandmothers' longevity negatively affects daughters' fertility. *American Journal of Physical Anthropology* 136: 223–229.

Maley, C. C., P. C. Galipeau, J. C. Finley, et al. 2006. Genetic clonal diversity predicts progression to esophageal adenocarcinoma. *Nature Genetics* 38: 468–473.

Mayer-Davis, E. J., S. L. Rifas-Shiman, L. Zhou, et al. 2006. Breast-feeding and risk for childhood obesity. *Diabetes Care* 29: 2231–2237.

Maynard, C. L., C. O. Elson, R. D. Hatton, et al. 2012. Reciprocal interactions of the intestinal microbiota and immune system. *Nature* 489: 231–241.

McKeown, T. 1976. *The Modern Rise of Population.* Academic Press, New York.

McKeown, T. 1983. Food, infection and population. *The Journal of Interdisciplinary History* 14: 227.

McNeill, W. 1976. *Plagues and People.* Random House, New York.

Meader, E., M. J. Mayer, D. Steverding, et al. 2013. Evaluation of bacteriophage therapy to control *Clostridium difficile* and toxin production in an in vitro human colon model system. *Anaerobe* 22: 25–30.

Medawar, P. B. 1952. *An Unsolved Problem of Biology*. H. K. Lewis, London.

Medzhitov, R., D. S. Schneider, and M. P. Soares. 2012. Disease tolerance as a defense strategy. *Science* 335: 936–941.

Mellon, M., C. Benbrook, and K. Benbrook. 2001. Hogging it. Estimates of antimicrobial abuse in livestock. Union of Concerned Scientists Publications, Washington, DC.

Metzger, M. W., and T. W. McDade. 2010. Breastfeeding as obesity prevention in the United States: A sibling difference model. *American Journal of Human Biology* 22: 291–296.

Meyer, U. A., and U. M. Zanger. 1997. Molecular mechanisms of genetic polymorphisms of drug metabolism. *Annual Review of Pharmacology and Toxicology* 37: 269–296.

Meyer, U. A., U. M. Zanger, and M. Schwab. 2013. Omics and drug response. *Annual Review of Pharmacology and Toxicology* 53: 475–502.

Milot, E., F. M. Mayer, D. H. Nussey, et al. 2011. Evidence for evolution in response to natural selection in a contemporary human population. *Proceedings of the National Academy of Sciences, USA* 108: 17040–17045.

Mischke, M., and T. Plosch. 2013. More than just a gut instinct—The potential interplay between a baby's nutrition, its gut microbiome and the epigenome. *AJP: Regulatory, Integrative and Comparative Physiology* 304: R1065–R1069.

Mogasale, V., B. Maskery, R. L. Ochiai, et al. 2014. Burden of typhoid fever in low-income and middle-income countries: A systematic, literature-based update with risk-factor adjustment. *Lancet Global Health* 2: E570–E580.

Moorad, J. 2013. A demographic transition altered the strength of selection for fitness and age-specific survival and fertility in a 19th century American population. *Evolution* 67: 1622–1634.

Moxon, R., C. Bayliss, and D. Hood. 2006. Bacterial contingency loci: The role of simple sequence DNA repeats in bacterial adaptation. *Annual Review of Genetics* 40: 307–333.

Muehlenbein, M. P., and R. G. Bribiescas. 2005. Testosterone-mediated immune functions and male life histories. *American Journal of Human Biology* 17: 527–558.

Murphy, K. P., P. Travers, M. Walport, et al. 2008. *Janeway's Immunobiology*. Garland Science, New York.

Murray, M. J., and B. A. Lessey. 1999. Embryo implantation and tumor metastasis: Common pathways of invasion and angiogenesis. *Seminars in Reproductive Endocrinology* 17: 275–290.

Neel, J. V. 1962. Diabetes mellitus: A "thrifty" genotype rendered detrimental by "progress"? *American Journal of Human Genetics* 14: 353–362.

Nees, F., S. H. Witt, A. Lourdusamy, et al. 2013. Genetic risk for nicotine dependence in the cholinergic system and activation of the brain reward system in healthy adolescents. *Neuropsychopharmacology* 38: 2081–2089.

Neff, F., D. Flores-Dominguez, D. P. Ryan, et al. 2013. Rapamycin extends murine lifespan but has limited effects on aging. *Journal of Clinical Investigation* 123: 3272–3291.

Nesse, R. M. 1994. An evolutionary perspective on substance abuse. *Ethology and Sociobiology* 15: 339–348.

Nesse, R. M. 2001. The smoke detector principle. Natural selection and the regulation of defensive responses. *Annals of the New York Academy of Sciences* 935: 75–85.

Nesse, R. M. 2005. Natural selection and the regulation of defenses. *Evolution and Human Behavior* 26: 88–105.

Nesse, R. M. 2015. Evolutionary psychology and mental health. In *Handbook of Evolutionary Psychology*, D. Buss (ed.), Wiley, Hoboken, NJ.

Nesse, R. M., and K. C. Berridge. 1997. Psychoactive drug use in evolutionary perspective. *Science* 278: 63–66.

Newman, R. A. 1988. Adaptive plasticity in development of *Scaphiopus couchii* tadpoles in desert ponds. *Evolution* 42: 774–783.

Ober, C., S. Elias, D. D. Kostyu, et al. 1992. Decreased fecundability in Hutterite couples sharing HLA-DR. *American Journal of Human Genetics* 50: 6–14.

O'Bleness, M., V. B. Searles, A. Varki, et al. 2012. Evolution of genetic and genomic features unique to the human lineage. *Nature Reviews Genetics* 13: 853–866.

Olsen, C. M. 2011. Natural rewards, neuroplasticity, and non-drug addictions. *Neuropharmacology* 61: 1109–1122.

Omer, S. B., D. A. Salmon, W. A. Orenstein, et al. 2009. Vaccine refusal, mandatory immunization, and the risks of vaccine-preventable diseases. *New England Journal of Medicine* 360: 1981–1988.

Omran, A. R. 2005. The epidemiologic transition: A theory of the epidemiology of population change. *The Milbank Quarterly* 83: 731–757.

Oria, R. B., P. D. Patrick, H. Zhang, et al. 2005. APOE4 protects the cognitive development in children with heavy diarrhea burdens in northeast Brazil. *Pediatric Research* 57: 310–316.

Pandrea, I., D. L. Sodora, G. Silvestri, et al. 2008. Into the wild: Simian immunodeficiency virus (SIV) infection in natural hosts. *Trends in Immunology* 29: 419–428.

Panter-Brick, C. 1996. Proximate determinants of birth seasonality and conception failure in Nepal. *Population Studies* 50: 203–220.

Parashar, U. D., E. G. Hummelman, J. S. Bresee, et al. 2003. Global illness and deaths caused by rotavirus disease in children. *Emerging Infectious Diseases* 9: 565–572.

Pardoll, D., and C. Drake. 2012. Immunotherapy earns its spot in the ranks of cancer therapy. *Journal of Experimental Medicine* 209: 201–209.

Parkhill, J. 2008. Whole-genome analysis of pathogen evolution. In *Evolution in Health and Disease*, S. C. Stearns and J. C. Koella (eds.), Oxford University Press, Oxford and New York, pp. 199–214.

Partridge, L., and N. H. Barton. 1993. Optimality, mutation and the evolution of ageing. *Nature* 362: 305–311.

Raberg, L., D. Sim, and A. F. Read. 2007. Disentangling genetic variation for resistance and tolerance to infectious diseases in animals. *Science* 318: 812–814.

Read, A. F., T. Day, and S. Huijben. 2011. The evolution of drug resistance and the curious orthodoxy of aggressive chemotherapy. *Proceedings of the National Academy of Sciences, USA* 108(Suppl. 2): 10871–10877.

Read, A. F., P. A. Lynch, and M. B. Thomas. 2009. How to make evolution-proof insecticides for malaria control. *PLoS Biology* 7: 1–10.

Reardon, S. 2014. Phage therapy gets revitalized. *Nature* 510: 15–16.

Redfield, R. J. 1988. Evolution of bacterial transformation: Is sex with dead cells ever better than no sex at all? *Genetics* 119: 213–221.

Risch, H., J. McLaughlin, D. Cole, et al. 2006. Population BRCA1 and BRCA2 mutation frequencies and cancer penetrances : A kin–cohort study in Ontario, Canada. *Journal of the National Cancer Institute* 98: 1694–1706.

Robertson, T. L., H. Kato, G. G. Rhoads, et al. 1977. Epidemiologic studies of coronary heart disease and stroke in Japanese men living in Japan, Hawaii and California. Incidence of myocardial infarction and death from coronary heart disease. *The American Journal of Cardiology* 39: 239–243.

Robison, A. J., and E. J. Nestler. 2011. Transcriptional and epigenetic mechanisms of addiction. *Nature Reviews Neuroscience* 12: 623–637.

Rogers, L. M., S. Veeramani, and G. J. Weiner. 2014. Complement in monoclonal antibody therapy of cancer. *Immunologic Research* 59: 1–8.

Rook, G. A. W. 2012. Hygiene hypothesis and autoimmune diseases. *Clinical Reviews in Allergy and Immunology* 42: 5–15.

Roseboom, T., J. van der Meulen, A. Ravelli, et al. 2001. Effects of prenatal exposure to the Dutch famine on adult disease in later life: An overview. *Molecular and Cellular Endocrinology* 185: 93–98.

Rosenberg, K., and W. Trevathan. 2002. Birth, obstetrics and human evolution. *BJOG: An International Journal of Obstetrics and Gynaecology* 109: 1199–1206.

Ross-Gillespie, A., M. Weigert, S. P. Brown, et al. 2014. Gallium-mediated siderophore quenching as an evolutionarily robust antibacterial treatment. *Evolution, Medicine, and Public Health* 2014: 18–29.

Roulette, C. J., H. Mann, B. M. Kemp, et al. 2014. Tobacco use vs. helminths in Congo basin hunter-gatherers: Self-medication in humans? *Evolution and Human Behavior* 35: 397–407.

Rushton, J. P. 2000. *Race, Evolution, and Behavior: A Life History Perspective*. 3rd edition. Charles Darwin Research Institute, Port Huron, MI.

Ryman, N., R. Chakraborty, and M. Nei. 1983. Differences in the relative distribution of human-gene diversity between electrophoretic and red and white cell antigen loci. *Human Heredity* 33: 93–102.

Sabin, A., W. A. Hennessen, and J. Winsser. 1954. Studies on variants of poliomyelitis virus. 1. Experimental segregation and properties of avirulent variants of three immunologic types. *Journal of Experimental Medicine* 99: 551–576.

Sanderson, S., G. Salanti, and J. Higgins. 2007. Joint effect on the N-acetyltransferase 1 and 2 (NAT1 and NAT2) genes and smoking on bladder carcinogenesis: A literature-based systematic HuGE review and evidence synthesis. *American Journal of Epidemiology* 166: 741–751.

Shanley, D. P., R. Sear, R. Mace, et al. 2007. Testing evolutionary theories of menopause. *Proceedings of the Royal Society of London, Series B, Biological Sciences* 274: 2943–2949.

Silva, A. S., and R. A. Gatenby. 2010. A theoretical quantitative model for evolution of cancer chemotherapy resistance. *Biology Direct* 5: 25–42.

Silva, A. S., Y. Kam, Z. P. Khin, et al. 2012. Evolutionary approaches to prolong progression-free survival in breast cancer. *Cancer Research* 72: 6362–6370.

Simantov, R., and S. H. Snyder. 1976. Morphine-like peptides in mammalian brain: Isolation, structure elucidation, and interactions with the opiate receptor. *Proceedings of the National Academy of Sciences, USA* 73: 2515–2519.

Smith, K. R., H. A. Hanson, S. S. Buys, et al. 2011. Effects of BRCA1 and BRCA2 mutations on female fertility. *Proceedings of the Royal Society of London, Series B, Biological Sciences* 279: 1389–1395.

Sommer, M. O. A., G. Dantas, and G. M. Church. 2009. Functional characterization of the antibiotic resistance reservoir in the human microflora. *Science* 325: 1128–1131.

Sørensen, M., H. Autrup, A. Olsen, et al. 2008. Prospective study of *NAT1* and *NAT2* polymorphisms, tobacco smoking and meat consumption and risk of colorectal cancer. *Cancer Letters* 266: 186–193.

Sottoriva, A., I. Spiteri, S. G. M. Piccirillo, et al. 2013. Intratumor heterogeneity in human glioblastoma reflects cancer evolutionary dynamics. *Proceedings of the National Academy of Sciences, USA* 110: 4009–4014.

Stearns, S. C. 1992. *The Evolution of Life Histories.* Oxford University Press, Oxford.

Stearns, S. C. 2012. Evolutionary medicine: Its scope, interest and potential. *Proceedings of the Royal Society of London, Series B, Biological Sciences* 279: 4305–4321.

Stearns, S. C., and R. E. Crandall. 1981. Quantitative predictions of delayed maturity. *Evolution* 35: 455–463.

Stearns, S. C., and D. Ebert. 2001. Evolution in health and disease: Work in progress. *Quarterly Review of Biology* 76: 417–432.

Stearns, S. C., and R. F. Hoekstra. 2005. *Evolution: An Introduction.* Oxford University Press, Oxford and New York.

Stearns, S. C., and J. C. Koella. 1986. The evolution of phenotypic plasticity in life-history traits: Predictions of reaction norms for age and size at maturity. *Evolution* 40: 893–913.

Stearns, S. C., and L. Partridge. 2001. The genetics of aging in *Drosophila*. In *Handbook of Aging*, 5th edition, E. Masoro, and S. Austad (eds.), Academic Press, Burlington, MA, pp. 345–360.

Stearns, S. C., N. Allal, and R. Mace. 2008. Chapter 3: Life history theory and human development. In *Foundations of Evolutionary Psychology*, C. K. Crawford (ed.), Lawrence Erlbaum Associates, New York, pp. 47–69.

Stearns, S. C., S. G. Byars, D. R. Govindaraju, et al. 2010. Measuring selection in contemporary human populations. *Nature Reviews Genetics* 11: 611–622.

Stewart, E. J., R. Madden, G. Paul, et al. 2005. Aging and death in an organism that reproduces by morphologically symmetric division. *PLoS Biology* 3: e45.

Strassmann, B. 1999. Menstrual cycling and breast cancer: An evolutionary perspective. *Journal of Women's Health* 8: 193–202.

Szreter, S. 2004. Industrialization and health. *British Medical Bulletin* 69: 75–86.

Tanner, J. M., and P. S. W. Davies. 1985. Clinical longitudinal velocities for height and height velocity for north American children. *The Journal of Pediatrics* 107: 317–329.

Tariq, S. M., S. M. Matthews, E. A. Hakim, et al. 1998. The prevalence of and risk factors for atopy in early childhood: A whole population birth cohort study. *Journal of Allergy and Clinical Immunology* 101: 587–593.

Tate, J. E., A. H. Burton, C. Boschi-Pinto, et al. 2012. 2008 estimate of worldwide rotavirus-associated mortality in children younger than 5 years before the introduction of universal rotavirus vaccination programmes: A systematic review and meta-analysis. *Lancet Infectious Diseases* 12: 136–141.

Teo, C.-G., and S. A. Locarnini. 2010. Potential threat of drug-resistant and vaccine-escape HBV mutants to public health. *Antiviral Therapy* 15: 445–449.

Thavagnanam, S., J. Fleming, A. Bromley, et al. 2008. A meta-analysis of the association between Caesarean section and childhood asthma. *Clinical and Experimental Allergy* 38: 629–633.

Thomas, F., A. T. Teriokhin, E. V. Budilova, et al. 2004. Human birthweight evolution across contrasting environments. *Journal of Evolutionary Biology* 17: 542–553.

Thompson, M. E., J. H. Jones, A. E. Pusey, et al. 2007. Aging and fertility patterns in wild chimpanzees provide insights into the evolution of menopause. *Current Biology* 17: 2150–2156.

Tian, T., S. Olson, J. M. Whitacre, et al. 2011. The origins of cancer robustness and evolvability. *Integrative Biology* 3: 17–30.

Tishkoff, S. A., F. A. Reed, A. Ranciaro, et al. 2006. Convergent adaptation of human lactase persistence in Africa and Europe. *Nature* 39: 31–40.

Trasande, L., J. Blustein, M. Liu, et al. 2012. Infant antibiotic exposures and early-life body mass. *International Journal of Obesity* 37: 16–23.

Trivers, R. L. 1974. Parent-offspring conflict. *American Zoologist* 14: 249–264.

Tsonos, J., D. Vandenheuvel, Y. Briers, et al. 2014. Hurdles in bacteriophage therapy: Deconstructing the parameters. *Veterinary Microbiology* 171: 460–469.

Tyner, S. D., S. Venkatachalam, J. Choi, et al. 2002. p53 mutant mice that display early ageing-associated phenotypes. *Nature* 415: 45–53.

U.S. Cancer Statistics Working Group. 2003. *United States Cancer Statistics: Web-based Report*. U.S. Department of Health and Human Services, Centers for Disease Control and Prevention, and National Cancer Institute; Atlanta. Available at: www.cdc.gov/uscs.

van Noordwijk, A. J., and G. de Jong. 1986. Acquisition and allocation of resources: Their influence on variation in life history tactics. *The American Naturalist* 128: 137–142.

Venning, G. R. 1983. Identification of adverse reactions to new drugs. II. How were 18 important adverse reactions discovered and with what delays? *British Medical Journal* 286: 289–292.

Viertel, T. M., K. Ritter, and H.-P. Horz. 2014. Viruses versus bacteria—Novel approaches to phage therapy as a tool against multidrug-resistant pathogens. *The Journal of Antimicrobial Chemotherapy* 69: 2326–2336.

Villalba, M., M. G. Rathore, N. Lopez-Royuela, et al. 2013. From tumor cell metabolism to tumor immune escape. *The International Journal of Biochemistry and Cell Biology* 45: 106–113.

Vivier, E., D. H. Raulet, A. Moretta, et al. 2011. Innate or adaptive immunity? The example of natural killer cells. *Science* 331: 44–49.

Wade, G., and J. Jones. 2004. Neuroendocrinology of nutritional infertility. *American Journal of Physiology: Regulatory, Integrative, and Comparative Physiology* 287: R1277–R1296.

Wagner, G. P., and V. J. Lynch. 2005. Molecular evolution of evolutionary novelties: The vagina and uterus of therian mammals. *Journal of Experimental Zoology Part B, Molecular and Developmental Evolution* 304: 580–592.

Wagner, G. P., and V. J. Lynch. 2010. Evolutionary novelties. *Current Biology* 20: R48–52.

Wang, X., S. G. Byars, and S. C. Stearns. 2013. Genetic links between post-reproductive lifespan and family size in Framingham. *Evolution, Medicine, and Public Health* 2013: 241–253.

Wang, Z.-X., J.-X. Cao, M. Wang, et al. 2014. Adoptive cellular immunotherapy for the treatment of patients with breast cancer: A meta-analysis. *Cytotherapy* 16: 934–945.

Weibel, E. R., C. R. Taylor, and L. Bolis. 1998. *Principles of Animal Design*. Cambridge University Press, Cambridge.

Weismann, A. 1882. *Ueber die Dauer des Lebens*. Verlag von Gustav Fischer, Jena, Germany, p. 94.

West, S. A., S. P. Diggle, A. Buckling, et al. 2007. The social lives of microbes. *Annual Review of Ecology Evolution and Systematics* 38: 53–77.

Wildman, D. E., C. Chen, O. Erez, et al. 2006. Evolution of the mammalian placenta revealed by phylogenetic analysis. *Proceedings of the National Academy of Sciences, USA* 103: 3203–3208.

Wilkins, A. S. 2002. *The Evolution of Developmental Pathways*. Sinauer, Sunderland, MA.

Williams, G. C. 1957. Pleiotropy, natural selection, and the evolution of senescence. *Evolution* 11: 398–411.

Williams, G. C. 1966. *Adaptation and Natural Selection*. Princeton University Press, NJ.

Wingo, P., C. Cardinez, S. Landis, et al. 2003. Long-term trends in cancer mortality in the United States, 1930–1998. *Cancer* 97: 3133–3275.

Winter, K., K. Harriman, J. Zipprich, et al. 2012. California pertussis epidemic, 2010. *The Journal of Pediatrics* 161: 1091–1096.

Wood, J. 1994. Menarche and menopause. In *Dynamics of Human Reproduction: Biology, Biometry, Demography*, J. Wood, Aldine de Gruyter, Hawthorne, NY, pp. 418–421.

World Health Organization. 2013. World Health Statistics 2013. Geneva, Switzerland.

Wright, G. J., and J. C. Rayner. 2014. *Plasmodium falciparum* erythrocyte invasion: Combining function with immune evasion. *PLoS Pathogens* 10: e1003943.

Wrigley, E., and R. Schofield. 1981. *The Population History of England 1541–1871; A Reconstruction.* Oxford: Blackwell, Oxford.

Wysowski, D. K., and L. Swartz. 2005. Adverse drug event surveillance and drug withdrawals in the United States, 1969–2002: The importance of reporting suspected reactions. *Archives of Internal Medicine* 165: 1363–1369.

Xie, H. G., R. B. Kim, A. J. Wood, et al. 2001. Molecular basis of ethnic differences in drug disposition and response. *Annual Review of Pharmacology and Toxicology* 41: 815–850.

Yachida, S., S. Jones, I. Bozic, et al. 2010. Distant metastasis occurs late during the genetic evolution of pancreatic cancer. *Nature* 467: 1114–1117.

Yates, L. R., and P. J. Campbell. 2012. Evolution of the cancer genome. *Nature Reviews Genetics* 13: 795–806.

Zaccone, P., Z. Fehervari, J. M. Phillips, et al. 2006. Parasitic worms and inflammatory diseases. *Parasite Immunology* 28: 515–523.

Index

Entries with an f next to the page number indicate that the information will be found in a figure. Entries with a t next to the page number indicate that the information will be found in a table.